Scrap Tires: Disposal and Reuse

Robert H. Snyder

Society of Automotive Engineers, Inc.
Warrendale, Pa.

> **Library of Congress**
> **Cataloging-in-Publication Data**
>
> Snyder, Robert H.
>   Scrap tires : disposal and reuse / Robert H. Snyder.
>   p.   cm.
>   Includes bibliographical references (p.   ) and index.
>   ISBN 1-56091-682-6 (alk. paper)
>   1. Waste tires--Management. I. Title.
> TD797.7.S64 1998
> 678'.32'0286--dc21   97-46942
>   CIP

Copyright © 1998   Society of Automotive Engineers, Inc.
                    400 Commonwealth Drive
                    Warrendale, PA 15096-0001 U.S.A.
                    Phone: (724) 776-4841
                    Fax: (724) 776-5760
                    http://www.sae.org

ISBN 1-56091-682-6

All rights reserved. Printed in the United States of America.

Permission to photocopy for internal or personal use, or the internal or personal use of specific clients, is granted by SAE for libraries and other users registered with the Copyright Clearance Center (CCC), provided that the base fee of $.50 per page is paid directly to CCC, 222 Rosewood Dr., Danvers, MA 01923. Special requests should be addressed to the SAE Publications Group. 1-56091-682-6/98 $.50.

SAE Order No. R-158

Table of Contents

Preface vii

Introduction ix
References xvi

Chapter 1 **The Problem of Scrap Tires** 1
Description of the Problem 1
Tires as a Fire Hazard 2
Health Hazards from Scrap Tires 3
The Origins of Scrap Tires 4
References 8

Chapter 2 **The Scrap Tire Collection Process** 9
Role of the Tire Jockey 9
The Dealer-Jockey Relationship 10
Disposition of Medium-Truck Tires 14
Dual Nature of the Scrap Tire Problem 15
Reference 18

Chapter 3 **Tire Processing and Its Problems** 19
Difficulties in Chopping Tires 19
Primary Tire Chopping 20
Screening and Sorting of Tire Chips 28
Utility of Coarsely Chopped Tire Chips 29
References 30

Scrap Tires: Disposal and Reuse

Chapter 4 **Further Comminution of Tire Chips** 31
  Particle Size as Related to Process 31
  Problems with Tire Cord and Tire Wire 32
  Ambient Grinding 34
  Cryogenic Grinding 35
  Wet Grinding 37
  Reference 37

Chapter 5 **Engineering Properties and Value of Tire Chips** 39
  Unique Properties of Scrap Tires 39
  Cost Considerations 41
  Perception of the Problem 42
  Hierarchy of Uses for Scrap Tires 44
  Reference 45

Chapter 6 **Fuel Uses of Scrap Tires** 47
  Characteristics of Tires Burned as Fuel 47
  Whole Tires as Boiler Fuels 50
  TDF as Supplemental Coal Fuel 51
  TDF as Wood Supplement 52
  TDF in the Portland Cement Industry 53
  Reference 56

Chapter 7 **Transportation Uses of Scrap Tires** 57
  Rubber in Asphalt 57
  Crumb Rubber in Crack Sealants 61
  Crumb Rubber in Repair Membranes 61
  Crumb Rubber in Paving Courses 62
  ISTEA Mandate on Use of Crumb Rubber
    in Highway Construction 65
  Rubber in Railroad Crossings 67
  References 70

Table of Contents

Chapter 8 Mats, Playturf, and Equestrian Uses of Scrap Tires 71
Rubber Mats 71
Agrimats 73
Rubber Mats for Playground Surfaces 74
Playturf from Scrap Tires 76
Rubber in Equestrian Arenas 78
References 80

Chapter 9 Scrap Tires in Sewage Sludge Composting and Soil Amendments 81
Sewage Treatment Plant Sludge 81
Tire Chips Replace Wood Chips in Composting ... 82
Effects of Crumb Rubber on Soil 86
Crumb Rubber as a Soil Amendment 87
References 93

Chapter 10 Civil Engineering Studies and Applications 95
Use as a Lightweight Aggregate in Fill 95
Use as Subgrade Thermal Insulation 98
Use as Backfill Behind Retaining Walls 98
Use in Landfills 98
Use in Septic Fields 99
Recent Problems in Using Tire Chips
  in Highway Construction 100
References 105

Chapter 11 Tire Pyrolysis 107
Description of the Process 107
Feasibility of Tire Pyrolysis 108
References 111

Chapter 12 Other Solutions: Fishing Reefs and Molded Rubber-Plastic Blends 113
Scrap Tires in Fishing Reefs 113
Molded Rubber-Plastic Blends 114
References 120

Chapter 13 Scrap Tire Regulations in the United States 121
Perception of a National Problem 121
The Need for State Regulation 121
Patterns Among State Regulations 123
Funding 124
New Programs 125
References 126

Chapter 14 Overview and Projections for the Future 127
Overview of Current Usage 127
Conclusions and Projections 130
References 134

Index 135

About the Author 149

Preface

In the early 1970s before any general recognition was given to a national scrap tire problem, the situation was brought to my attention as part of another problem.

A lieutenant on the Detroit police force had written a long letter to the president of the U.S. Rubber Company about a problem that was plaguing the police, namely, abandoned vehicles. Every month the Detroit police coped with approximately 2,000 vehicles that had been abandoned and had subsequently become a police problem. The police officers worked with the local car junkies with good success except for two items: scrap tires and upholstery. The lieutenant's letter, which asked if we could help him, came to my desk, with the admonition of "Please handle." Even 30 years later, I am still trying to "handle" the problem.

In response to the letter, I spent the next few days in the lieutenant's company, visiting the depots where the abandoned vehicles were sequestered. The vehicles were kept for six months while the police tried to identify the owners and determine whether the vehicles had been used in a crime. Assuming no answers were found, the vehicles were now ready for the auto junkies, with whom the police had regular dealings. We visited several sites that were typical, admiring the expertise that goes into separating the more valuable residual metal components from those of less value and finally coming face to face with the large piles of scrap upholstery and the scrap tires that remained mounted on wheels because of high dismounting costs. At that time, the

Scrap Tires: Disposal and Reuse

mounted tires were likely to be hauled to a dump and burned, but public opinion was rising in opposition to the practice. The auto upholstery was mostly Saran, originally chosen because it was nonflammable; however, at this point, the upholstery was cursed with a high scrap cost because it could not be disposed of by burning.

This event prompted my interest in the scrap tire problem, an interest that grew steadily and persisted into my retirement. More recently, I am happy to acknowledge a debt to Norman Emanuel of Baltimore, Maryland, the guru of all tire choppers in terms of depth of experience and technical expertise, for his friendship and advice during the past 30 years. I also extend gratitude to Louie and Tim Baker of Baker Rubber. Finally, I express my appreciation for the splendid support that I received from Kuivi, my gracious and lovely wife for 52 years, and from our children, one of whom developed my affliction and does business as JaiTire Industries, chopping tires in Denver and marketing Crown III nationally as a turfgrass top dressing.

Robert H. Snyder
February 1997

Introduction

From the beginning of the automotive industry, vehicles have been equipped with rubber tires. It can be argued that the modern automobile would not have been possible without Robert Dunlop's invention of the pneumatic tire and its enormous subsequent development.

In the 1920s, the average mileage that could be expected from a pneumatic tire was only a few thousand miles, in contrast to the 40,000- to 50,000-mile treadlife expectancy of today's tires. During the 1920s, tires also were much larger. Modern textiles and improved rubber compounds have permitted substantial increases in the load-carrying capabilities of tires.

During most of the history of the automobile, a scrap tire problem did not exist. The problem began in the early 1960s and has become acute only during the last 15 years. It is a matter of interest and some relevance to understand why this occurred.

Prior to World War II, tires were constructed from cotton textile cords, natural rubber, and (then as now) steel wire beads. Natural rubber, as received from the Indies, came in 200-lb bales of highly elastic but nonplastic smoked sheets which are intractable in the rubber factory. Before articles can be shaped and molded from raw natural rubber, the rubber first must undergo substantial processing to reduce it to a limp, plastic state that is suitable for the calendering and extrusion operations required to assemble the raw tire before vulcanization.

Scrap Tires: Disposal and Reuse

Calendering is the process in which extended sheets of uniform films are produced by squeezing rubber or plastic between large counterrotating rolls. In extrusion, thick shaped pieces are prepared by forcing the material through a die under pressure from a rotating screw. Vulcanization is the process in which sulfur is combined with rubber or plastic in a chemical operation that links the polymer chains, thereby increasing strength, stability, and elasticity of the polymer. These preparative operations were expensive and time consuming. Fortunately, the processing of natural rubber compounds could be expedited by the use of substantial amounts of previously vulcanized but reclaimed rubber compounds. Here, reclamation refers to a process by which a vulcanized rubber object is devulcanized or otherwise degraded back to a plastic mass.

One of the properties of natural rubber compounds is its ability to slowly devulcanize when heated strongly. In the jargon of the compounder, the natural rubber compounds are said to "revert." For tires or other rubber articles in service, this property clearly is a defect; however, this defect becomes an advantage in rubber reclaiming. In the devulcanization process, what appears to occur is that some of the polymeric chains in the rubber polymer molecule are cleaved, and some of the sulfur cross links are broken. The net effect results in smaller, modified rubber molecules that can be recompounded and revulcanized to produce rubber articles with reasonably good properties, although not as good as those of articles made from virgin rubber.

More specifically, this devulcanized rubber compound (reclaimed rubber) can be added in substantial amounts to new natural rubber to yield compounds of superior processability and minimal loss of quality. In the face of such obvious value, reclaiming processes were developed to accelerate the reaction and to

Introduction

produce reclaimed rubber of high value at low cost. Prior to the period following World War II, such tires as could be collected economically were collected and reclaimed in hundreds of facilities throughout the United States. All of the major U.S. tire companies operated reclaiming plants. According to annual reports provided by the U.S. Department of Commerce,[I-1] 291,082 long tons of reclaimed rubber were used in 1943, and as late as 1960 reclaim consumption was 278,703 long tons.

The unavailability of natural rubber during World War II necessitated the development of synthetic rubber. Initially, this new synthetic rubber was of poor quality and intolerable processability. At the end of World War II, synthetic rubber remained generally inferior to natural rubber. Melvin Mooney, one of the great figures in rubber science and technology, once wrote a sober comparison of the technical quality of early styrene-butadiene rubber (SBR) versus natural rubber and concluded that "the properties of SBR bore the same relation to those of natural rubber that a hen's egg does to that of a stork!" He was referring primarily to treadwear resistance and cut and tear resistance of compounds of that day. But with continuing development, the true virtues of SBR in terms of treadwear superiority and the absence of the reversion defect began to emerge, and synthetic rubber slowly supplanted natural rubber in tires.

Of particular importance in the development of synthetic rubber was the commercial benefit of long-range price stability of the synthetic product in comparison to the wild price swings that occurred with natural rubber. Several times within a decade, the price of natural rubber fluctuated from $0.05/lb to more than $1/lb during one calendar year. These wild price swings in a major commodity whose markets were 10,000 miles distant were intolerable. There were several years when at least one major

U.S. tire manufacturer lost more money in rubber inventories than it made in operations.

In the continued technical development of synthetic rubber, polymer chemists achieved great control of the process and were able to produce SBR of superior processability, which could be further improved by the addition of cheap petroleum oils at prices of approximately $0.04/lb, in contrast to reclaim at $0.14/lb. Inevitably, rubber reclaiming became a dying industry. In 1996, no commercial operation was reclaiming tire rubber domestically, and the single remaining operator, U.S. Rubber Reclaiming Company, reclaims only butyl tubes at the level of approximately 1,500 tons annually.[1-2]

In tracking the demise of the previously prosperous rubber reclaiming business, one other important point must be understood. Even in the heyday of the industry, it was impossible to use reclaim in amounts greater than 20% without serious loss of quality. Tires made entirely from reclaim would have been of commercially unsatisfactory quality. Accordingly, it was never possible to reclaim all of the tires discarded in any year and to reuse them by putting them back into that same number of new tires to be produced the following year (i.e., 100% recycling of tires back to tires). However, during the thriving years of the rubber reclaiming industry, many articles made from rubber that could also use reclaim are no longer manufactured from rubber today. Modern plastics have replaced rubber in many large-volume uses. Wire insulation, floor tile, floor mats, garden hose, and other articles were once made chiefly of rubber compounds, and those articles also could employ large amounts of reclaimed rubber. In recent times, polyethylene or polyvinyl chloride (PVC) has replaced rubber in many of those items. Thus, the corresponding markets for reclaim have disappeared.

Introduction

As noted above, the level of use of reclaim in tires has declined drastically. In addition to the technical reasons, one important contributing factor was the decline in the use of snow tires. Years ago, the average car owner maintained a pair of snow tires, which had to be mounted on the car for the winter and removed in the spring. Often, these snow tires were retreads, and the rubber compounds used in their bulky treads usually contained substantial amounts of reclaim. All-season tires have virtually displaced conventional snow tires in the marketplace, with a considerable diminution in reclaim usage.

Finally, the adoption of radial tires has played a role in the demise of reclaim markets. The greater tread life of radial tires and enhanced service requirements, such as low rolling resistance, have created a situation where the compounder must use recipes providing highly controlled sets of properties from virgin materials of narrow product variability.

In summary, the decline of the rubber reclaiming industry was the major cause of today's scrap tire problem. If all 278,000 tons of reclaim used in 1960 had been produced from scrap tires, more than 50 million scrap tires would have been required. Other rubber articles also were being reclaimed in 1960. Therefore, not all reclaim was derived from tires, although tires were the chief component.

An important secondary cause of the present scrap tire problem is the recent and continuing decline in the passenger tire retreading industry. Years ago, one of every four worn-out tires was retreaded. As late as 1970, 35 million passenger tires were retreaded annually, compared to new passenger tire sales of 169 million in that year, including tires supplied on new automobiles sold that year. In recent years, that number has dropped significantly. The following table cites figures from the International Tire and Rubber Association (formerly the American Retreaders Association) for passenger tires and medium-truck tires for the years 1982 to 1997.

TIRES RETREADED (MILLIONS)*

Year	Passenger	Medium Truck
1982	25.5	—
1985	15.5	13.1
1988	15.1	14.1
1991	8.4	14.8
1993	6.6	15.4
1994	5.9	15.9
1995	5.0	16.0
1996	4.2	16.5
1997	3.9 (Est.)	16.7 (Est.)

*Courtesy of Marvin Bozarth, International Tire and Rubber Association (formerly American Retreaders Association)

Truck tire retreading actually increased during this period. Large truck tires, as opposed to medium, are of low volume and do not merit discussion here.

Retreading a tire does not eliminate it from contributing to the scrap tire problem because the retreaded tire will eventually wear out, usually in somewhat less time than a new tire. However, retreading defers the appearance of those casings on the scrap heap and thus materially reduces the number of scrap tires that would have appeared if they had not been retreaded.

Three reasons can be cited as chief contributors to the causes for the steep decline in passenger tire retreading. The first factor, as

Introduction

already noted, was the decline in the use of snow tires, a high proportion of hich were retreads.

The second factor was the appearance of steel-belted radial ply tires, which presented two problems. By their design and construction, radial ply tires generally could not be retreaded using processes and equipment that were used to retread bias ply tires. Major investments in equipment and in molds would be necessary to become a successful retreader of radial ply tires. Furthermore, the quality of the steel-belted radial ply tires supplied by many domestic manufacturers at the outset was not particularly suitable for retreading, and the performance level of retreaded tires from early radial tire production was frequently unsatisfactory. Obviously, the original casing would have traveled two to three times as far as its bias ply counterpart would have traveled and thus had a correspondingly high exposure to weathering, to sidewall damage, to fatigue, and to punctures. The yield of casings from the field that could be successfully retreaded was greatly reduced. Together with the equipment problem, the net effect was to eliminate many retreaders from the business who could not endure the financial strain of the transition period.

The third factor to damage the passenger tire retreading business was and is the importing into the United States of large numbers of low-cost, new radial ply tires, chiefly from the Asian basin. These tires generally are of satisfactory quality for replacement tires, have the perceived advantage of being new tires, and are priced at levels close to the price of quality retreaded casings.

The decline in sales of retreaded casings had an impact on the scrap tire problem in excess of what could be inferred from the decline itself. To understand and appreciate this, we must look at the process and the routes by which scrap tires arrive at their ultimate resting place.

References

I-1 *Rubber Age*, Vol. 89, No. 2, p. 356, 1961.
I-2 LeGrone, Don, private communication, 1967.

Chapter 1

The Problem of Scrap Tires

Description of the Problem

The large and rapidly growing accumulation of piles of scrap vehicle tires throughout the United States exists chiefly because of a lack of markets for the scrap tires. This situation must be remedied. To compound the problem, scrap tires are undesirable landfill components because of their low bulk density, their tendency to float up to the surface of the landfill, and the fact that they are a shifting and unstable base for future building construction.

Open above-ground piles of scrap tires constitute a problem for a number of reasons. First, they are bulky and unsightly, and they tend to accumulate in highly populated areas where they are conspicuous. More importantly, however, they are potential fire and health hazards. Both of these problems also stem in part from the bulky nature of scrap tires. Their toroidal (i.e., doughnut) shape tends to preclude nesting or close packing. Loose piles of tires have a bulk density of approximately 10 lb/ft^3.

Tires as a Fire Hazard

In regard to flammability, tires behave as though they are a mixture of jellied petroleum and carbon. They are not subject to spontaneous combustion and are not easily ignited. However, once afire, they burn vigorously. Their low bulk density, or loose packing, ensures ample access to air. If large piles of whole tires begin to burn, the fire is virtually inextinguishable.

In the open, tires burn with a hot, sooty, and malodorous flame. In doing so, they emit an entire spectrum of undesirable chemicals, including some carcinogens. In addition, the fierce heat of the fire produces considerable accompanying pyrolysis of adjacent tires in the pile. Pyrolysis describes chemical changes produced by the application of heat alone, as opposed to oxidation or burning. Substantial quantities of petroleum oil are produced and can create a runoff problem. Unchecked, this oil runoff can enter neighboring streams or percolate through the soil, thereby contaminating the groundwater.

In 1983, a serious tire fire occurred in Winchester, Virginia. Apparently, the fire was set by an arsonist. A pile of approximately 5 to 7 million tires, collected over several years, was stacked 80 ft deep at some points and extended over 4 acres. Once aflame, the fire was inextinguishable, in spite of several weeks of massive efforts by a large crew of experts. Because the oil runoff was well contained at that point, the decision was made to let the fire burn itself out. That took approximately three months. In the early stages, the oil runoff rate was measured at 6,000 gal/hr. It was entrained in ponds dug for that purpose which were lined with polyethylene. The runoff eventually came to approximately 800,000 gal, of which 700,000 gal were of merchantable quality and could be sold as fuel oil.

The Problem of Scrap Tires

The U.S. Environmental Protection Agency (EPA) estimated its costs at more than $1 million.[1-1]

A more recent, well-publicized tire fire occurred in 1990 in Hagersville, Ontario. This fire was partially contained by physically separating and removing from the pile large numbers of unburned tires. The fire was extinguished after 17 days. In this case, it was possible to sample the smoke and effluent gases and to document the presence of hundreds of chemical compounds, mainly hydrocarbons.

Apart from the high firefighting costs, the residue from the Hagersville fire presents a substantial cleanup problem. The ashes of the fire contain large amounts of steel, zinc oxide, carbon char, and oily residues in the soil. The oily residues represent the major environmental problem because they may be leached away and contaminate water sources in the process. However, none of them nor the mixture serves as a simple earthen soil replacement, and thus soil must be replaced.

Health Hazards from Scrap Tires

At least two well-recognized health problems are associated with scrap tires. Again, both problems arise from the loose structure of random tire piles. First, this type of structure presents an elegant apartment house for vermin, inaccessible to their predators. If a food source can be found in the vicinity, tire piles may become infested with rats.

A more significant health problem exists with regard to mosquitoes. Regardless of how a tire is placed in a pile, its shape will serve as a convenient receptacle to catch rainwater. The average tire in a pile can contain more than 1 qt of water after a heavy rain. These pools of water, which are protected from sunlight, serve as ideal breeding grounds for mosquitoes. Cases of encephalitis have been traced to specific colonies in tire piles. The Lacrosse encephalitis strain was named after the Wisconsin tire pile from which it was isolated and identified.

Mosquito infestation of tire piles appears to be a global problem. Tire casings imported from Asia have contained mosquito larvae, which matured into adults that have escaped and have introduced a more virulent strain into the U.S. mosquito problem.

Apart from the above health hazards, scrap tire piles are not believed to pose a threat to the environment. In spite of frequent invitations to take a stand on the matter, EPA has not identified scrap tires as a hazardous substance. Indeed, EPA approved the use of crumb rubber as a playground surface and as a soil amendment.[1-2]

The Origins of Scrap Tires

The circumstances of the sale and use of vehicle tires afford a simple and exact method to quantify them as they arise and to identify their location.

Tires serve their useful life as component parts of vehicles. They were designed for use on those vehicles, and their use is largely confined to the vehicles for which they were designed. When the tires on those vehicles wear out, one of two things must occur. The tire must be replaced with another tire, usually a new tire, or the vehicle and its tires must be scrapped.

The Problem of Scrap Tires

It follows that one scrap tire arises for every new tire that is sold, of the same kind and size. Moreover, the scrap tires arises at the location where the replacement tire is sold. Detailed and reliable figures are available for sales of replacement tires. These figures will account for approximately 80% of all passenger scrap tires.

At this point, note that tire retreading, to whatever extent practiced, does not interfere with the validity or rigor of these facts. Retreading is the most valuable thing to do with worn-out tires that have been removed from a vehicle because it delays the appearance of that tire on the scrap heap. However, retreading does not eliminate the ultimate appearance of that tire on the scrap heap, nor does it change the number of scrap tires.

Years ago, one of every four worn tires was retreaded. As late as 1970, 35 million tires were retreaded compared to new passenger tire sales of 169 million in that year, including tires supplied on new cars that year.

The other path by which tires enter the scrap heap is less obvious and often overlooked. Every time an automobile is scrapped, four or five scrap tires arise. This is the simple fact. Any usable tires were replaced before the vehicle was scrapped. There is a way to count these tires. First, recognize that the annual increase in vehicle registrations is small—approximately 3%. Accordingly, each year we scrap almost as many vehicles as are sold during that year. Obviously, the annual sale of tires to vehicle manufacturers (OE sales) closely relates to the number of vehicles that are scrapped by automotive recyclers. For example, if 10 million new passenger vehicles are sold in a given year, automotive recyclers will generate more than 40 million scrap tires. These constitute approximately 20% of the total. Together with those scrap tires arising at tire dealers, they account for 99% of all scrap tires. Table 1.1 lists recent U.S. sales figures for OE and replacement passenger tires, as provided by the Rubber Manufacturers Association.

TABLE 1.1
TOTAL DOMESTIC TIRE SHIPMENTS*
(THOUSANDS OF UNITS)

Year	Original Equipment	Replacement	Total
Passenger Tires			
1981	35,979	125,263	169,242
1985	54,839	141,455	196,294
1990	47,199	152,251	199,450
1993	52,335	165,148	217,483
1994	58,448	169,983	228,431
1995	56,965	166,844	223,809
1996	57,092	175,328	236,420
Light-Truck Tires			
1981	2,600	17,000	19,600
1985	4,000	19,800	23,800
1990	3,712	22,832	25,544
1993	4,593	23,805	28,398
1994	5,889	25,351	31,240
1995	6,042	25,537	31,579
1996	5,834	27,605	33,439
Medium-Truck Tires			
1983	2,394	10,407	12,801
1985	3,580	9,752	13,332
1990	3,062	10,113	13,175
1993	3,792	10,361	14,153
1994	4,850	11,072	15,922
1995	5,427	10,740	16,167
1996	4,273	10,723	14,996

*Courtesy of the Rubber Manufacturers Association, 1400 K St., N.W., Washington, DC.

Another significant fact about scrap tires is that they are "source separated" at the point where they are generated. Most users do not individually dispose of their tires, as they do with their newspapers and bottles. Source separation will greatly simplify the ultimate solution of the scrap tire problem, but it is of little value after scrap tires are collected into piles.

To return to the question of the magnitude of the scrap tire problem, the size of the problem is obviously the total number of tires that have been produced minus the number that have been adequately disposed.

The shape of the problem is another matter. First, distribution is uneven from a geographical standpoint. Obviously, tires are worn out by the people and industries that use them. Therefore, the worn-out tires arise where the people and industries are located, and to a good first approximation at the rate of one passenger tire per person per year. Treadwear rates are strongly influenced by geography and road construction, but the main controlling factor of scrap tire generation is demography. The population chart of the United States, broken down into states and counties, is a reliable and frequently used guide to projected tire sales and thus to scrap tire generation. The main point is that scrap tires are anchored to their original location, chiefly by their high cost of transportation. As we have noted previously, tires have a low bulk density. A 40-ft semitrailer can hold approximately 1,000 to 1,200 25-lb tires, depending on how carefully they are packed. This creates a total load of 25,000 to 30,000 lb, far less than its load-carrying capability. Tires resist handling by conventional material handling techniques. They usually are loaded individually to save space or, less effectively, on pallets. Generally, it costs approximately one dollar to transport a new tire 100 miles. Scrap tires occupy almost as much space as they did when they were new and thus have approximately the same cost to ship. Therefore, scrap tires tend to remain close to where they were generated.

In addition, what are now or were recently considered to be acceptable means of disposal are key factors in determining the severity of a scrap tire problem. Landfilling tires was considered an acceptable means of disposal until small but heavily populated states such as New Jersey began running out of landfill area. In many regions of the United States, tires were burned with other rubbish in dumps until Clean Air Act legislation prohibited the practice. In Pennsylvania, enormous numbers of tires were dropped down the mineshafts of abandoned mines. In Los Angeles, the LaBrea Tar Pits accepted large volumes of tires without complaint. In Connecticut, the U.S. Rubber Company operated one of the largest and last tire reclaiming plants in the United States. Closure of that plant precipitated a scrap tire problem over the entire state of Connecticut. Thus, the available means of disposing of scrap tires directly affects the magnitude of the scrap tire problem.

References

1-1 *EPA Environmental News*, U.S. Environmental Protection Agency, February 21, 1994.
1-2 U.S. Environmental Protection Agency, Comprehensive Guidelines for Procurement of Products Containing Recovered Materials (CPG) (60 FR 21370), May 1, 1995.

Chapter 2

The Scrap Tire Collection Process

Role of the Tire Jockey

As explained in Chapter 1, most scrap tires and other worn tires arise first at tire dealers' locations. What happens to these tires next is key to understanding the whole process. The tire dealer must dispose of the worn-out tires promptly. A large tire dealer may collect 1,000 tires in one week. Each time a dealer sells a tire, that dealer acquires a used casing of almost the same volume. These used casings make a bulky pile, and the dealer has no room to store them. The dealer's business is selling new tires; therefore, he basically has no interest in the used casings except to remove them from his property. They are in his way.

Typically, a new player now enters the scene—a "tire jockey" who, for a fee, periodically picks up the dealer's used casings and transports them to his place of business.

Among the used casings collected by the dealer are a significant fraction that are not worn out and thus can be used either as secondhand tires or as retreadable tires. Considerable money can be made by sorting from the used casings those that have genuine value for resale in the proper channels.

The basic definition of a worn-out tire is one that has less than 2/32 in. of remaining tread depth. Most tire users have their own less-discriminating standard of determining when a tire is worn out. They often prefer to replace a complete set of tires at one time, regardless of unequal wear among the tires. Contemporary front-wheel drive cars wear front tires more rapidly than tires on the rear wheels. Unless the owner has practiced careful tire rotation, it is likely that when a whole set of tires is replaced, some of those tires will have significant remaining usable tread depth and the others may continue to meet standards for retreadability. The tire jockey's role clearly is valuable to the driving public and should not be abrogated. Note that the problems of inspection, sorting, distribution, and sale of these used tires require a specialized knowledge that is not usually available to tire dealers.

Tire jockeys operate in every metropolitan area, and virtually all dealers use their services. Tire jockeys may differ widely in the scope and mode of their operations, depending on local conditions and increasingly on state tire regulations aimed at controlling the collection and transport of scrap tires.

The Dealer-Jockey Relationship

To see in more detail how the dealer-jockey relationship works, let us consider a hypothetical case. In this scenario, the tire jockey visits the tire store every Friday and picks up a load of 1,000 tires. The tire jockey also collects payment from the dealer. The payment amount can vary depending on the location, but in this case let us assume that the payment is $0.50 per tire. This usually is called the tipping fee. For this payment, the tire jockey takes these 1,000 tires to his place of business and sorts them. The tire jockey removes 180 tires that can be resold as secondhand tires at $3 per tire, and 120 tires that can be sold to a retreader at $2 per casing. The tire jockey's gross income position is as follows:

From the dealer	1,000 tires at $0.50/tire	$ 500
From the tire sale	180 tires at $3/tire	$ 540
From the retreader	120 tires at $2/tire	$ 240
Gross income		$1,280

On the negative side, the tire jockey also has 700 genuine scrap tires of which he must dispose. How he disposes of these scrap tires is critical to remaining in business.

A few years ago, the tire jockey may have been able to take the scrap tires to a nearby landfill for a low fee. However, landfills now discriminate against tires as being too bulky, as providing an unstable subterranean structure that is unsuitable for building construction, and as having a tendency to float to the surface after several years, buoyed by gases contained in the upper portion of the toroid. Today, the tire jockey would have to pay a substantial premium to dispose of his tires in a landfill, if he could even find one that would accept them. Therefore, in our scenario, what the tire jockey has been doing is hauling the scrap tires to a rural area. There the tire jockey's cousin has a farm with a deep ravine on the property, and the tire jockey has been dumping the scrap tires into the ravine. Now the ravine is full, and the tire jockey must drive 100 miles farther across the state line to dump his tires. Obviously, this substantially increases the tire jockey's expenses. The tire jockey's only recourse is to raise his price to the dealer, assuming that the retreader will not pay more. Repeated from time to time, this process provokes the expected reaction from the dealer, who looks for a cheaper way to be rid of the used tires. Enter another new player, a "gypsy tire jockey," who undercuts the normal price. This gypsy tire jockey has access to a big abandoned building located downtown, where he can surreptitiously dump scrap tires.

This scenario, in endless small variations, illustrates how the game was and, to some extent, continues to be played. Every scrap tire must be taken someplace! Prohibition of one or another mode of disposal does not solve the problem and ensures that another, perhaps even more odious, disposal method probably will arise.

The scene continues to change, however. As discussed in future chapters, the tire jockey frequently can take scrap tires to a nearby cement kiln where they are burned for fuel, or to a local steam generation plant where they can be used as a cheap supplement to the normal coal boiler fuel.

Following along these lines, some tire jockeys or their associates occasionally have had convenient local arrangements that legally permitted them to inconspicuously pile large quantities of tires. The biggest known pile, which was located in Modesto, California (see Figure 2.1), probably contained 20 million tires before a power plant was built on the site to use the tires as a low-cost fuel source. In or near most metropolitan areas, large piles containing millions of scrap tires exist. The owner of such a tire pile is likely to view the pile as a future source of wealth when uses for those tires will be found. The owner may make token efforts to chop the tires for fuel or other uses. He probably is impecunious and cannot afford the cost of removing the tire pile or breaking it into smaller piles that could be managed in the event of a

The Scrap Tire Collection Process

Fig. 2.1. Tire pile in Modesto, California, containing 20 million tires.

fire. The owner received payments for the tires in that pile, but those payments were far less than the expense that would be required to eliminate the pile or its associated hazards. Frequently, the taxes on the land have not been paid, and the property has reverted to the local government. If enough pressure is brought to bear, that tire pile may happen to catch fire one night!

Disposition of Medium-Truck Tires

Before concluding this portion of our discussion, note that what has been said applies chiefly to the disposition of passenger tires and light-truck tires. Medium-truck tires, which are defined as 10.00–20 and larger sizes that are used on over-the-highway trucks and semitrailers, constitute an exception. Small-fleet owners or owners of individual trucks may buy tires through a local dealer, but large trucking fleets have different marketing arrangements and their scrap tires may travel to the scrap pile by a slightly different route. Although their numbers (see Table 1.1 in Chapter 1) are small, a medium-truck tire weighs 100 lb or more, four times as much as a passenger tire. This weight fraction becomes more significant in the disposal problem. Medium-truck tires will not be uniformly distributed; rather, they will tend to remain a concentrated but isolated portion of the scrap tire stream. We have already noted that many more truck tires than passenger tires are retreaded. A retreadable truck tire casing has a market value of $30 or more, significantly more than a passenger tire, which has a value of approximately $2.50. Relatively few truck tires will escape retreading, with many being retreaded two or more times. On the other hand, medium-truck tires are more difficult to chop and thus command a higher tipping fee. Many truck tires are monoply steel construction instead of textiles, and they have much thicker wire beads than passenger tires. Most tire chopping equipment used for passenger tires is unsuitable for truck tires. Therefore, truck tires tend to be left for last when scrap tire piles are chopped. However, when truck tires are their way into the tire cavity and, if undetected, can enter the tire chopper, severely damaging the equipment.

5. On the other hand, tires received directly from a dealer or a tire jockey are clean. When chopped, these clean tires can meet more rigorous specifications regarding dirt content in their product output. Chapter 8 shows that some high-value crumb rubber products have tight specifications for dirt content.

The above discussions clearly demonstrate that, for convenience and for cost reasons, a tire recycler will prefer to receive tires directly from a tire jockey rather than from a long-standing tire pile. This conclusion is of profound significance with regard to the public need to remove existing piles of scrap tires. Where prompt removal is required, a heavy subsidy or cost penalty to accomplish this will be needed. No tire recycler will willingly collect those dirty tires at his own expense until the supply of fresh, clean scrap tires is exhausted. That is the bad news. The good news is that as capacity is put into place to deal with the ongoing stream of new scrap tires, the old piles of scrap tires will not loom as large. Present opinion[2-1] is that the number of scrap tire piles that represent an imminent threat to public health or safety does not contain more than approximately three times the annual rate of scrap tire disposal. As time passes, these can be handled as a secondary matter—*en passant*, so to speak. Projections about the full resolution of the scrap tire problem will be discussed in a later chapter.

Reference

2-1 "1996 Scrap Tire Market/Disposal Study," Scrap Tire Management Council, 1400 K St. N.W., Washington, DC 20005, April 1997.

Scrap Tires: Disposal and Reuse

between huge piles of scrap tires and scrap tires close to the point of origin become clear.

1. A successful business enterprise must be based on the availability of a steady, ongoing stream of scrap tires and knowledge of the size of that stream. It would seem foolish to make a substantial capital investment in a plant to chop an existing pile of tires, lacking a stable, ongoing supply of freshly discarded tires in the appropriate quantity to continue operations. However, if that ongoing supply of tires is available, it will be preferred to the tire pile unless the cost of processing the existing pile can be heavily subsidized.

2. Because of the awkward handling characteristics of tires, retrieval of tires from a scrap tire pile is expensive. The scrap tires must be individually loaded onto a truck and transported to the processing site, where they next must be laboriously unloaded. Unless several trucks are engaged in this effort, significant idle time at the tire pile will result because of awaiting return of the truck.

3. If tires go to the tire chopping plant directly from the dealer or the tire jockey, one whole transportation and loading episode can be avoided and the cost of the remaining one can be shared.

4. Tires brought from an outdoor tire pile will contain water in the summer and several pounds of ice in the winter. In addition, these tires are dirty and often are mixed with other scrap materials. They may contain dirt as a result of being bulldozed along the ground or through mud. This earthen dirt contains sand, which is extremely abrasive to the steel in the tire chopper. Furthermore, pieces of steel scrap often find

chopped, they yield approximately four times as much ground crumb rubber as a passenger tire. The concept of Passenger Tire Equivalents (PTE) has evolved to estimate the yield of ground rubber from a mixture of truck and passenger tires.

Dual Nature of the Scrap Tire Problem

One useful conclusion that can be inferred from these discussions is that there is not only one scrap tire problem. Rather, two separate and distinct problems exist.

The first problem is the present outdoor, above-ground, uncovered accumulation of scrap tires. The second problem is the ongoing generation of scrap tires before their aggregation into large collections. Both problems require a solution, but clearly they require entirely different treatment. This becomes more obvious after consideration of the following.

One tacit assumption underlying all our discussions is that a solution to the nation's scrap tire problem does exists, and that solution involves the establishment of profitable commercial enterprises and markets to utilize scrap tires for their inherent intrinsic value.

As we shall subsequently see, whole scrap tires may be burned as a practical energy source, but this is a low-value use. Higher-value products require that tires be chopped into narrow distributions of particle sizes but in a number of different size ranges. Thus, a second assumption is that scrap tires will require chopping and size grading, and often will require removal of the wire, if not the textile fabric. In light of these assumptions, some of the more significant value differences

CHAPTER 3

Tire Processing and Its Problems

Difficulties in Chopping Tires

Although there has been and continues to be some use of whole tires (passenger tires and medium-truck tires) as fuel (*vide infra*), as a practical matter any other effectual use of tires requires them to be chopped into reasonably uniform pieces in a size range appropriate to their intended use. However, this need has been the main obstacle in the utilization of scrap tires. For at least two reasons, scrap tires are difficult to chop.

First, this difficulty occurs because the rubber and the wire, each in its own way, resist cutting. Tire wire is a medium carbon steel effectively hardened by several cold drawing steps to a Rockwell hardness of approximately 65. Machine edges hard enough to cut the wire repetitively tend to be brittle and to chip. Softer edges require frequent and usually expensive replacement. Rubber compounds also are difficult to cut because frictional forces build quickly on the sides of the blade when the knife edge begins to penetrate the rubber. Anyone who has ever attempted to use a knife to cut through a 1-in. thickness of a piece of rubber knows that it is extremely difficult unless you stretch the rubber to prevent it from binding on the knife and wet the knife to reduce friction.

Second, tire rubber compounds are abrasive. This unobvious fact is the worst problem. Heavy frictional contact with a piece of tire rubber quickly dulls the sharpest steel edges. In tire manufacturing operations, even the knives that cut soft, unvulcanized rubber need frequent sharpening and are kept wet during operations to reduce friction and wear. One manifestation of the abrasiveness of tire treads can be seen on examination of a well-traveled highway. Approximately 95% of the exposed surface appears to be smooth, flat stones. These stones were not flat in the original aggregate. They have been polished flat by tire wear.

As a consequence, commercially available equipment that is used to chop tin cans, scrap steel, refrigerators, or steel barrels is virtually useless for chopping tires. The lengthy list of purveyors of chopping equipment who attempted to adapt their equipment to chop tires and then abandoned the business should serve as a warning.

Primary Tire Chopping

As we continue, it will become clear that the choice of tire chopping equipment is heavily influenced by the particle size of the intended product and other specification requirements. Often, if not usually, several successive chopping or grinding operations must be conducted to arrive at the final product, and different pieces of equipment may be necessary. At this time, however, we will confine our remarks to the primary passenger or light-truck tire chopping operation whereby the whole tire is chopped into pieces ranging from 10 to 25 in^2.

Today, the most popular machines consist of arrays of spaced circular blades on pairs of counter-rotating shafts. Opposite each blade on one shaft, a spacer of equal width on the other shaft permits the blades to counter-rotate in overlapping fashion. One or more hooks may project from the blade periphery. These do not cut but merely drag the tire into the knives. Ideally, the knife blades operate with small clearances in the overlap region (i.e., 0.005 in.). The cutting edges are the corners of the knives which, together with the adjoining blade edges of the knives on the opposing shaft, provide the necessary scissors action. Figure 3.1 is a schematic drawing of a pair of counter-rotating blades. In practice, a battery of 20 or more of these blades, together with alternating spacers, may be mounted on each shaft, which is hexagonal in the drawing but in some machines is keyed to lock the

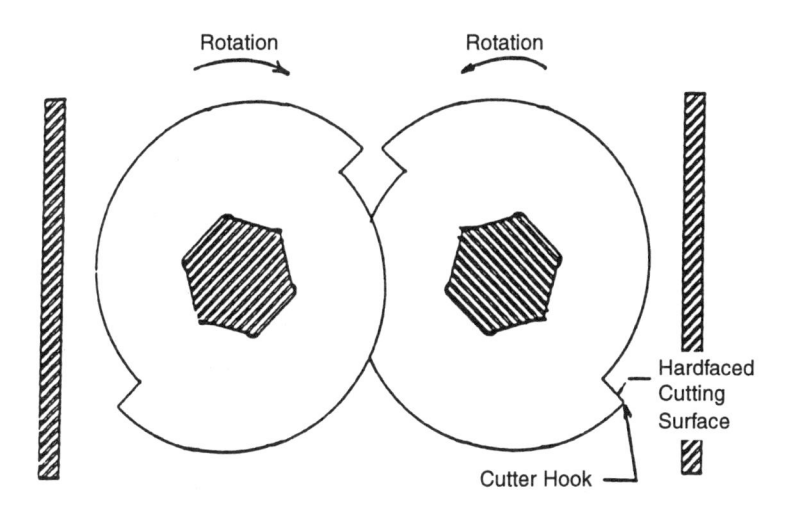

Fig. 3.1. Rotary shear shredder with one-piece counter-rotating knives. (Courtesy of Columbus McKinnon Corporation)

knives into fixed positions. Large locking nuts are threaded onto the ends of the shafts to lock the knives and spacers firmly into place.

Machines of this type are called hook and shear shredders, originally designed in the 1960s by Kurt Rosler, a German engineer. These machines are widely used for size reduction of scrap items ranging from steel drums to refrigerators. When used to shred tires, a single passage of the tire through the blade assembly would not produce chips but instead would produce hoop-shaped strips of tire sections of varying lengths and of the width of the blade cutters. Typically, the strips would be recycled through the blades and recut into chips of a size that can pass through a sifting device below the knives and then out of the machine.

If the machine has a new set of blades with sharp knife edges and close clearances between the opposing blades, it will produce cleanly cut chips at minimal power consumption. After some period of operation, however, the knife edges will begin to wear and will become rounded. At that point, rubber will begin to be dragged between the blades. The chips now start to show exposed lengths of tire cords or wires that are not cut but instead are torn from the tire. The rubber trapped between the blades also produces effective braking action, increasing the power consumption of the machine. The rubber trapped between the blades abrades more metal, permitting more rubber to be trapped and requiring more power until, inevitably, the required torque becomes excessive and the machine stalls. In modern machines, excessive power consumption triggers a clutch that reverses the direction of blade rotation to free the jammed rubber, and the machine starts forward again. As blade wear progresses, the frequency of these events increases and productivity decreases.

At that point, it is prudent to stop operations and then remove and resharpen the knives.

Blade sharpening is not a trivial or inexpensive procedure. Both sides of each blade must be ground on the entire blade surface until sufficient metal has been removed. Because the blade now is narrower than it was prior to grinding, the opposing spacer also must be ground to the same thickness as the blade. This must be done with precision machine tools, and machinists are necessary. Blade sharpening is expensive. Blades may be resharpened several times but not indefinitely because considerable metal is removed at each sharpening. After several sharpenings, what was once a 2-in. blade may be only 1.5 in. wide.

In view of this situation, it is not surprising that blade costs are the main maintenance expense encountered. They may easily amount to as much as $10 per ton of chopped product. To reduce blade sharpening costs, several ingenious blade modifications are being practiced. These include use of a series of blade widths on the same machine which downsize one size after each sharpening until discarded, or split blades whose positions on the shaft can be reversed to present a fresh, sharp edge without resharpening.

A machine similar to hook and shear shredders is the Holman shredder, which was designed specifically to shred steel-belted tires. (See Figure 3.2.) This machine has detachable blade sections bolted to a knife rotor. The rotor spacers are permanently attached to the knife rotors, and wear plates also are secured to each side of the rotors to protect against wear. (See Figure 3.3.) The machine design permits closer spacing tolerance between

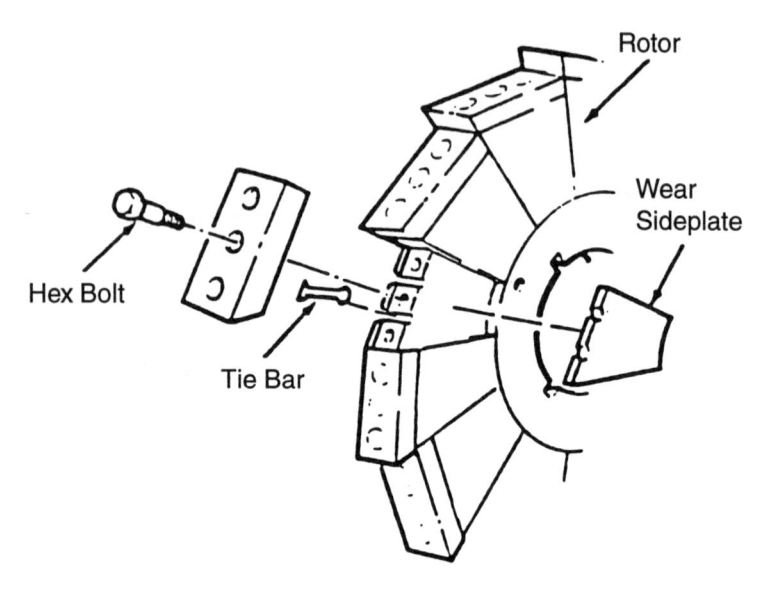

Fig. 3.2. Detail of Holman shredder.
(Courtesy of Columbus McKinnon Corporation)

the edges of adjacent blades than can be achieved with conventional hook and shear machines. The smaller detachable blades can be made of harder, more wear-resistant steel alloys. On the other hand, no hooks exist to grasp the tire, and therefore feed rollers are necessary to pinch the tire and urge it into the bite of the knives. Columbus McKinnon Corporation holds the patent rights to the Holman shredder.

Several single-rotor machines are available on the market, on which shaped cutter blades pass through matching fixed stator blades. After resharpening, the metal loss can be accommodated by repositioning the stator blades closer to the rotor blades, permitting multiple resharpenings before blade replacement becomes

Tire Processing and Its Problems

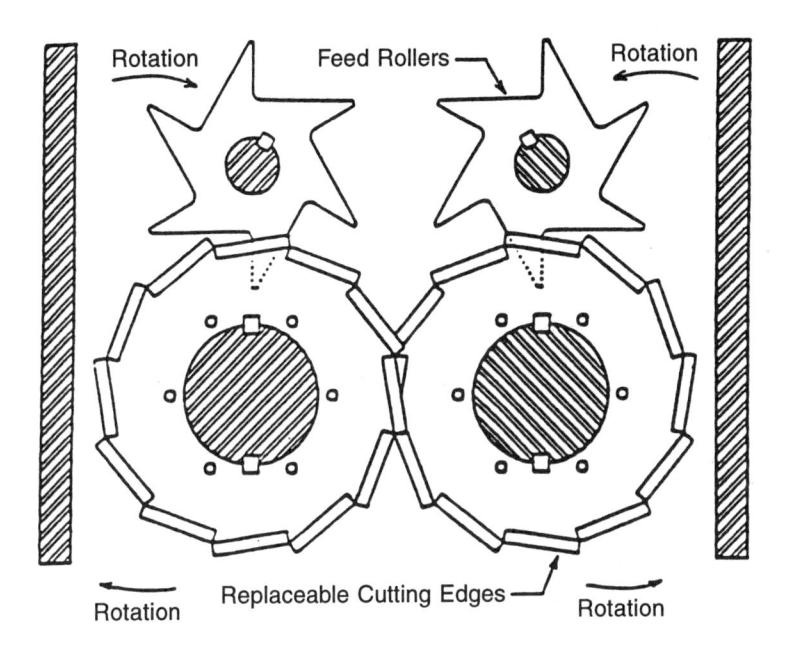

*Fig. 3.3. Rotary shear shredder with detachable knives.
(Courtesy of Columbus McKinnon Corporation)*

necessary. Several types of screw machines also have appeared of the scene,[3-1] but none has had a major commercial impact to date.

In addition, several commercial operations use hammermills as shown in Figure 3.4. Pounding on tires to dismember them may seem implausible and inefficient. However, one such machine, built by the Diamond Z Manufacturing Company[3-2] and originally designed to chop wood waste, is functioning effectively as a tire chopper in Traverse City, Michigan. This machine's virtues are that it can chop all kinds of tires including aircraft tires

Scrap Tires: Disposal and Reuse

Fig. 3.4. Diamond Z hammermill in operation in Traverse City, Michigan. (Courtesy of Diamond Z Manufacturing)

and the largest commercial off-road sizes, some 12 ft in diameter, reducing them to 3-in. chips. This machine does an excellent job of tearing the wire loose from the rubber, which then can be conveniently separated magnetically. The capacity is enormous—as much as 30 tons/hr. It is powered by two 800-HP caterpillar engines, and most of that energy is used to heat the rubber. A constant stream of water (25 gal/min) is required for cooling, and substantial amounts of steam are generated. Fuel consumption is high—approximately 30 gal/hr of diesel fuel for each motor. The machine also is noisy, but overall it is impressive.

A comprehensive, up-to-date list of suppliers of tire-chopping machines is available in the *Scrap Tire Users Directory*, published by Recycling Research Institute, 133 Mountain Road, Suffield, CT 06078.

Tire Processing and Its Problems

The cost of tire shredding equipment encompasses a broad range, depending on the ton-per-hour capacity and the performance characteristics required. One small, well-regarded machine called the Untha[3-3] (Figure 3.5), capable of chopping passenger tires to 1-in. chips at the rate of approximately 1 ton/hr, costs slightly less than $90,000. Popular machines with a capacity of more

*Fig. 3.5. Untha tire grinder and shredder.
(Courtesy of Tryco International)*

than 1,000 tires per hour (10 tons) can cost $300,000 to $600,000, depending on their features.

Before we leave this topic, it is worth reiterating that all pneumatic tires are difficult to chop, and truck tires are extremely difficult. In the eyes of the users of primary tire chopping machines, the best machines available today are only marginally satisfactory. Breakdowns occur frequently, and maintenance costs are high.

Screening and Sorting of Tire Chips

The output from a primary tire shredder usually requires some screening, if only to return grossly oversize pieces to be chopped again. The most basic configuration is a screen positioned at the base of the chopping chamber. Small pieces fall through the screen. The larger pieces are swept up to the top by the cutter blade hooks and are cut again. Some commercial machines utilize a large barrel-shaped screen (trommel), which rotates at a speed appropriate to hold the oversize pieces against the screen until they reach the top of the trommel. There, the oversize pieces drop into a conveyor and are returned to the shredding chamber. These configurations suffer from the disadvantage that tire chips with protruding wires become caught in the screen and tend to block it, necessitating manual removal.

Perhaps the most effective sorting and sizing device is the old-fashioned disc screen, similar to the type that was used to sort potatoes. Rows of rotating shafts with suitably spaced daisy wheels mounted on them sweep the oversize pieces into a return conveyor, allowing smaller pieces to fall through the spaces between the wheels. Disc screens may be used exterior to the tire shredder, or contained within it, as practiced by Columbus McKinnon Corporation.

Utility of Coarsely Chopped Tire Chips

Coarse chopping of tire chips effects approximately a threefold reduction in bulk volume and makes the tire chips more acceptable as landfill material. The coarse chopping also makes the tire chips less of a fire or health hazard if the chips are stored above ground. An interesting but somewhat controversial use for coarsely chopped tires is a low-density aggregate used in civil engineering applications. (This topic will be discussed in a later chapter.) Coarsely chopped tires also find limited use as a fuel or a supplementary fuel; however, in the latter case, the chopped tires must meet sizing requirements. In general, the particle size of the supplementary fuel must effectively correspond to the particle size of the primary fuel. Otherwise, if the rubber chips are larger, they will tend to accumulate, incompletely burned.

Usually the primary, coarse chips produced in the first stage must be chopped again and sorted into chips with a specific size distribution. As noted, the primary sorting device may be external to the chopper, or it may be self contained and restrict passage of oversize chips until they have been chopped again to a size that will pass through the screen. For adequate productivity, the specified chip size should be related to the blade spacing. If the blade separation is 2 in., a good yield of chips can pass through a 2-in. sieve. On the other hand, if we need 0.5-in. chips, a poor yield will result because chips intermediate in size between 0.5 in. and 2 in. will be recycled repetitively through the chopper with minimal size reduction. Efficient production of 0.5-in. chips requires a second-stage processor, perhaps of the same design as the primary chopper but with blades more closely spaced, preferably 0.5 in. It would be desirable to have a single, large-volume machine that could efficiently reduce the tire to 0.5-in chips in one stage. However, the design of such a machine is beyond our capabilities at this time, and the power requirements would be formidable. The relatively small Untha machine[3-3] appears to do this but at some productivity loss.

As a practical matter, production of 0.5-in chips or smaller usually requires two or more machines operating in tandem. In most cases, further size reduction will require different equipment.

References

3-1 U.S. Patent No. 5,115,988 to E.O. Tolonen, May 26, 1992.
3-2 Diamond Z Manufacturing, 1102 Franklin Boulevard-N, Nampa, ID 83687.
3-3 Untha Tire Grinder, Tryco International, Box 1277, Decatur, IL 62525-1277.

CHAPTER 4

Further Comminution of Tire Chips

Particle Size as Related to Process

The commercial markets for tire chips and crumb rubber comprise particle sizes ranging from approximately 1 in. down to 80 mesh. To achieve these size requirements, several different processes come into play. The upper range of particle sizes is best achieved by cutting processes, down to a size of approximately 0.25 in. Below this size, grinding processes are more effective and, as a matter of operating convenience, also may be used on large tire chips or tread buffings.

For intermediate rechopping of tire chips smaller than 1 in. down to approximately 10 to 12 mesh, granulators are emerging as the preferred machine. Granulators resemble reel-type lawnmowers in their action, as shown in Figure 4.1. A set of long, straight knives mounted on a common axle rotates at high speeds and at close clearances past a set of fixed bed knives chopping a stream of tire chips fed between them. The chopped mass is sifted continually to remove small particles. These may then be sifted again and sorted into appropriate size classifications.

Scrap Tires: Disposal and Reuse

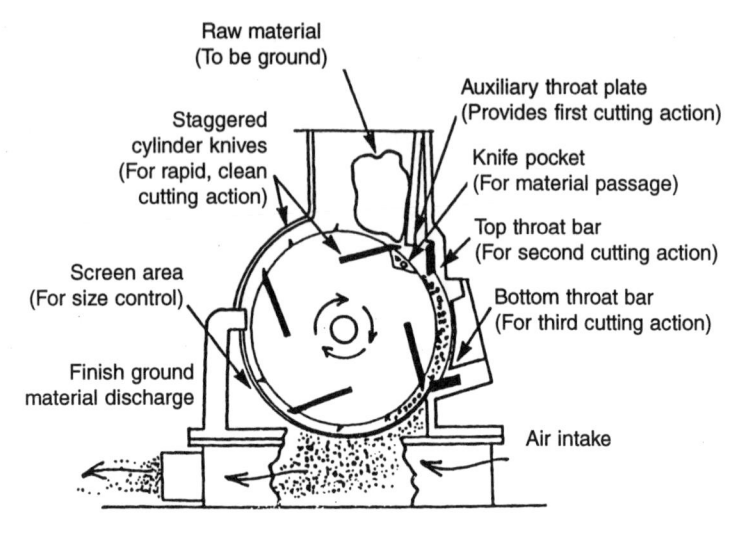

Fig. 4.1. Granulator mechanism.
(Courtesy of Columbus McKinnon Corporation)

Problems with Tire Cord and Tire Wire

Granulators do a good job on tread peelings or other pieces that are wholly rubber. However, production of rubber particles from tire chips presents two particular problems, namely, removal of the tire cord and removal of the tire wire. The tire cord (chiefly polyester and nylon) can be chopped more easily than the rubber. Consequently, the tire cords are quickly degraded to short lengths of individual filaments. These tiny filaments tend to aggregate into fluffy balls that float on the bed of rubber chips. The filaments are easily airborne and usually are collected in a cyclone collector, together with rubber fines. No important use has been found for these filaments to date; therefore, they usually are baled and discarded in a landfill. Ironically, the filaments could be considered as the best-looking and most uniform product of the operation; however, they are worthless.

Tire wire presents a different problem. Tire wire is tough enough to resist cutting by granulators. Tire wire may be of two kinds: bead and belt wire. Both types are medium carbon steel and after cold drawing are copper or bronze plated to permit good adhesion to the wire. The bead wire is used as a single filament usually 37 mils in diameter. The bead of a passenger tire can contain 20 to 25 turns of this wire. Beads are so damaging to granulator blades that they usually must be removed prior to the granulation process, either by debeading the tire before chopping or by magnetic separation of the coarse tire chips containing the beads. Belt wires are laid in helical cables, usually containing four to nine filaments of fine wire, 5 mils in size. The belt wire filaments largely escape chopping during the granulation process. Instead, the granulator blades shear the rubber off the wire, exposing a bright metallic surface. The belt wire then is collected magnetically as short crimped lengths of wire, with small quantities of rubber attached.

Recovered belt wire offers interesting potential as high-quality steel scrap, but it presents its own set of problems. The copper on the surface of the wire is objectionable in most steels. Fortunately, the amount of copper present in the scrap wire is low enough to be tolerable. Any remaining rubber attached to the wire also presents a problem because of the contained sulfur, which is highly objectionable in steel-making operations. Furthermore, the fine, resilient character of the wire resists baling and has low bulk density. If heated in air, oxidation will occur because of the large surface area. The wire is suitable for use only in electric furnaces. Some wire is now being sold to scrap dealers, but most is placed in landfills.

In normal operations with a granulator, the main product tends to be material in the 0.25-in. size. Small quantities of crumb rubber in the size range of 20 to 30 mesh will be produced, together with large particles. However, if the chief material to be produced is intended to be 20 mesh or finer, an alternate method to grinding must be found.

Ambient Grinding

Because of their elastic nature, rubber particles resist grinding even more than they resist cutting. However, several processes are being used commercially to grind rubber chips to fine particles. The most widely used process involves repetitive grinding of rubber chips between large-diameter iron refining mill rolls, often corrugated and run at uneven speed. (See Figure 4.2.) After being caught in the nip, the rubber chips are subject to a shearing force resulting from the differential peripheral roll speed. Because this occurs at room temperature, this process often is identified as ambient grinding.

Ambient grinding is applicable to any size particle, including the whole tire. The process was developed by tire reclaimers as a means of obtaining small particles suitable for digestion in reclaim heaters. With the decline of reclaim and the growing demand for crumb rubber, ambient grinding has come into its own as a process to make crumb rubber down to the 20 to 30 mesh range. Usually, depending on the size of the starting particle, a train of several mills (usually three) will be used. The first one may reduce the original article to small chips. The second in the chain will grind the chips to separate the rubber from the metal and fiber. Then a finishing mill will grind the material to the intended product specification. After each

Further Comminution of Tire Chips

Fig. 4.2. Refiner mill. (Courtesy of Baker Rubber, Inc.)

pass through a mill nip, the product is classified by sifting screens, returning oversize pieces to the mill nip.

Cryogenic Grinding

The so-called cryogenic process has developed slowly but is now becoming popular, particularly as a means of making particles in the 50 to 80 mesh size. In this process, whole tires first are comminuted to tire chips of approximately 3-in. size. These chips are then frozen using liquid nitrogen at -195°C (-319°F) and its cold vapors. The rubber freezes at approximately -75°C (-103°F). Freezing converts the tough elastic particles to a brittle, glassy state in which they are easily shattered (often with a hammermill) into tiny smooth-sided particles and are also separated from any adhering wire or fabric. In the frozen state, they are classified on shaker screens to the desired particle size and the removal of the fabric and wire.

The cryogenic process has received significant attention in recent years but until lately has not been a commercial success for scrap tires. The process is being practiced commercially on scrap of specialty rubbers such as butyl, paracril, and neoprene, which are ground to 50 to 80 mesh. The specialty rubbers often are suitable for recycling directly into the same virgin compound and then can be resold to the original producer. Success may rest on the fact that the specialty rubbers are more expensive and thus can support a higher process cost for recovery.

Cryogenic grinding appears to have several economic disadvantages, compared to ambient grinding. Economically, it is tied to the cost of liquid nitrogen. Because liquid nitrogen is difficult to transport, operations must be located conveniently close to the liquid air plant.

A major portion of the manufacturing cost resides in the price of liquid nitrogen and the amount needed to freeze the rubber chips. The price of liquid nitrogen is somewhat arbitrary and depends on alternative uses and markets. In recent years, that price has been in the range of $0.02 to $0.05/lb. In the 20 to 30 mesh range, ambient grinding was cheaper than cryogenic grinding. However, this disadvantage seems to have been overcome, and several new cryogenic facilities are said to be competitive in the 20 to 30 mesh range. The surface character of particles produced by cryogenic grinding differs significantly from the surface character of particles produced by ambient grinding. Cryogenic grinding involves conchoidal fracture of a brittle material to yield particles with smooth surfaces. In contrast, ambient grinding produces particles with rough, reticulated shapes and relatively high surface-to-volume ratios. The difference is either an advantage or a disadvantage, depending on the end use. For asphalt extension by most processes, ambient ground crumb rubber is preferred; for recycling into virgin rubber compounds, cryogenic material seems to be preferred.

Wet Grinding

In addition to conventional ambient grinding and the cryogenic process, several proprietary wet grinding processes are in limited use.[4-1] Tiny rubber chips immersed in water are ground between hard circular grinding plates moving countercurrently and lubricated by the water. When the appropriate particle size range is reached, the wet rubber is removed and dried. Productivity is relatively low; however, very small particle sizes can be obtained in the range of 400 to 500 mesh.

Reference

4.1 Rouse, Michael, private communication, Rouse Rubber, Box 831, Vicksburg, MS 37180, 1997.

CHAPTER 5

Engineering Properties and Value of Tire Chips

Unique Properties of Scrap Tires

At this point in our analysis of the scrap tire problem, it is clear that the means exist to reduce scrap tires to narrow distributions of uniformly sized chips or particles in sizes ranging from 2-in.^{2} pieces down to 450 mesh.

The major problem that we will now address is how to find profitable uses and applications for these products. The primary difficulty is that in each of the many particle size ranges, tire chips have somewhat different engineering properties than other materials in that same size range and in most instances do not qualify as direct replacements for materials to which they seem comparable. To make this point more obvious, Table 5.1 lists some of the more interesting properties of scrap tire particulates. This is truly a unique set of properties to be found in one material. However, compared to other more common materials, this combination of properties is both a blessing and a bane.

TABLE 5.1
ENGINEERING PROPERTIES OF RUBBER PARTICULATES

Black—Opaque
Liquid state—Low freezing point
Low density (1.15)
Water resistant—Nonwicking
Low thermal conductivity—Thermal barrier
Low electrical conductivity—Insulator
High absorption capabilities for most organics
Elastic, compliant, and resilient
Flammable
High heat of combustion—Low ash
Organic—Nonbiodegradable

For example, the hydrocarbon character and high heat of combustion of tire chips should make them an ideal solid fuel replacement for coal. However, the absence of high ash levels precludes their use as a total replacement for coal in a conventional boiler where the coal ash serves the necessary function of protecting the grate from the high combustion temperatures. Rubber chips are limited to approximately a 15% replacement level when used as a supplementary fuel to coal to avoid the problem of burned-out grates.

In a more dramatic example, Ingersoll Rand, a manufacturer of heavy earth-compaction equipment, maintains a track at its test headquarters on which to run and test its compactors. The track surface was stone gravel. Under the hard, heavy-footprint loading of the compactor, the gravel soon was reduced to powder and required frequent replacement. The test superintendent had the bright idea of using tire chips to replace the gravel and ordered one truckload as a trial to cover a small segment of the course. In his words, "We couldn't get the drivers off that portion of the course." At considerable cost, the test superintendent purchased enough rubber chips to cover the entire course. Years later, the surface remains in place, providing a dust-free road bed.

The expectation, or at least the hope, is that the unique combination of physical properties of rubber chips will find outlets in unique products. In later chapters, we shall see how this is happening.

Cost Considerations

An equally important consideration in the prospective utilization of tire chips or crumb rubber is the cost. Depending on the particular product under discussion, several different cost elements compose the total. These will be discussed in subsequent chapters. At the moment, however, it is useful to focus on two cost elements for special consideration: one intrinsic element and one extrinsic element.

First, the cost of a rubber particulate will depend on the amount of chopping or grinding necessary to achieve the desired particle size. In chopping chips, we are generating new surfaces. The fundamental cost of the operation will depend on the amount of cutting that we must do (i.e., the surface area that we generate). As we reduce the size of the rubber chips, we also are increasing the ratio of surface to volume, which increases process costs. In reducing 2-in. chips to 1-in. chips, the surface-to-volume ratio doubles. Progressive size reductions entail a similar doubling ratio, with corresponding cost increases.

Second, the cost of the product will be determined by the cost of the starting material (i.e., the cost of the scrap tire at the door of the recycling plant). Today, that cost is negative. The tire processor is paid to chop the tire. The amount of payment may vary significantly from one location to another—from as much as $5 per passenger tire in some urban locations to as little as $0.50 per passenger tire. In any case, the tipping fee probably represents the largest single cost element in the recycling scheme. This reflects the public anxiety and concern about our scrap tire problem. In many cases, the tipping fee is a direct public subsidy paid to relieve the problem.

The important point here is that the performance of the present system is not guaranteed. As late as 1986, one major private brand tire dealer was able to sell its worn-out tires, rather than pay a significant penalty charge to dispose of them. We can ask what circumstances could arise to revert the present situation back to the previous one. This probably will not happen within the next few years. However, the following analysis suggests that it can happen and should warn potential investors in scrap tire projects. The present tipping fee situation is a fragile one.

Perception of the Problem

Tires compose only a small percentage (1–2%) of the waste stream. Because tires are unsightly, objectionable, and identifiable, they occupy a more significant position in public perception than they deserve. If we examine the annual production figures for polyethylene and polypropylene,[5-1] we see that 26.518 billion lb of polyethylene and 11.991 billion lb of polypropylene were produced in 1996. If we compare those figures to those for an estimated 275 million tires for the same year, whose weight we would estimate at 25 lb per tire or 687 million lb total, it is clear that we produce five times as much polyethylene and polypropylene as we do tires. The scrap rate presumably is in the same ratio if we make the reasonable assumption that we do not accumulate large dissimilar amounts of either in the environment. Obviously, polyethylene and polypropylene together are larger factors in the waste stream by a factor of five and are equally nondegradable in a landfill. However, they are far less conspicuous and do not present the same fire or health hazards.

Likewise, consider only the number of tons of coal burned by utilities each year to generate electricity. If we burned only two scrap tires with every 5 tons of coal that is burned, we should burn all our scrap tires.

Finally, in a normal year, approximately 30 million tons of asphalt were used in building roads. Thus we see that the 3.5 million tons of passenger tire equivalents will replace only approximately 12% of the asphalt consumed if all our scrap tires were used for that purpose.

The point of all this is not to minimize the scrap tire problem or its importance, but to demonstrate that, in real terms, the problem is not as large as we imagine and to suggest that any of the attractive potential uses for tire chips could easily reach proportions to preempt the total available supply. In that case, the tipping fee would be endangered and the tire dealer might again be able to sell his scrap tires. The likelihood of this plainly depends on a high value use for tire chips or crumb rubber.

Scrap Tires: Disposal and Reuse

Hierarchy of Uses for Scrap Tires

As a final step in this analysis, let us consider the hierarchy of uses for scrap tires and their derivatives. At the top of this hierarchy is retreading—the best alternative for a worn-out tire is to retread it. For reasons already discussed, most worn-out tires cannot be retreaded. However, for those that can be retreaded, this represents the customer use of most value.

At the bottom of the hierarchy is landfilling of scrap tires. This is a negative value. The lowest positive value accrues to the use of scrap tires as fuel. Except for the steel and small amount of inorganic matter, the remainder of the tire can be considered as petroleum derived. (The natural rubber content of the tire is an agricultural product, but the synthetic natural rubber product also exists and is used.) Then the tire can be considered as a suspension of carbon black particles in jellied petroleum that has temporarily been diverted from the fuel use assigned to most petroleum. We could burn it and recover the full BTU value of petroleum, approximately 15,000 BTU/lb. Stated in these terms, the average passenger tire is equivalent to slightly more than 3 gal. of fuel oil, or approximately $3.50 (roughly $40/ton of tire). If the chopped tire is used as a coal replacement, the intrinsic value may be slightly less in some areas.

On the other hand, if the tire is used as an asphalt extender and we assign it the same value as the asphalt that it replaces, the tire becomes worth more than $100/ton. Alternatively, if we comminute it to fine particles and incorporate it into virgin rubber compounds, or blend it with certain plastics, the value can rise to $250 to $500/ton.

To put this in perspective, we can say (without getting ahead of our story) that the fastest and most accessible use for scrap tires is as fuel. However, several other potentially high-volume uses to be discussed in later chapters and that can demonstrate higher value will become established in the next few years. These uses will supplant the use of tires as fuel after the full supply of tires is being recycled.

Reference

5.1 *Chemical and Engineering News*, June 23, 1997, p. 43.

CHAPTER 6

Fuel Uses of Scrap Tires

Characteristics of Tires Burned as Fuel

To begin this chapter, I would like to clarify some popular misconceptions about the use of tires as fuel. If we ignite a tire in the open (or even a pile of tires), we have an extremely hot flame accompanied by a sooty, malodorous smoke containing sulfur dioxide and a catalog of hydrocarbons and other chemicals. So why can burned tires be considered as a good fuel? The undesirable consequences of open burning of tires arise primarily from incomplete combustion. Carbon black particles and many chemical pyrolysis products escape the combustion zone without being combusted. Thus, the remedy is a larger combustion zone, which a proper furnace can provide.

An analogy is that if we restrict the air supply to the flame of a gas torch, we see a luminous, sooty flame. The luminosity arises from escaping incandescent carbon particles. Given a better air supply, complete combustion is achieved as evidenced by a bright blue flame. In a proper boiler, tires or tire chips burn completely, emitting normal combustion gases but including small amounts of sulfur dioxide. Tires normally contain approximately 1.2% sulfur on average. When burned, some of this is converted to sulfur dioxide and emitted. However, tires actually contain significantly less sulfur than most common coals used as fuel. Midwestern soft coals often contain as much as 4% sulfur. Montana coals and other Western coals may possess lower sulfur levels (e.g., 0.5%), but they also produce fewer BTU per pound (e.g., 6,000 versus 13,500) than are obtained from tires and on an equivalent BTU basis are only marginally superior to tires as far as sulfur dioxide emissions are concerned.

In summary, tires can be an excellent fuel. When we consign tires to a landfill, we are paradoxically burying better fuel than we are mining. Table 6.1 lists typical combustion heats for common boiler fuels.

TABLE 6.1
HEAT OF COMBUSTION OF VARIOUS FUELS

Fuel	BTU/lb
Oak wood	8,300
Pine wood	9,100
Lignite	6,000
Bituminous coal	11,000–14,000
Fuel oil	18,000–19,000
Tire chips	13,000–14,000

Tires also contain significant amounts of iron and zinc, which usually are not present in coal. The zinc level in modern tires is approximately 1.5%. On combustion it meets a divided fate. Some zinc can be found in the ash, but some is entrained as a particulate in the stack gases where it may increase the opacity level if released. Normally, zinc would be contained in the baghouse or other collection device of a modern boiler, along with the fly ash.

Fuel Uses of Scrap Tires

Tires also contain larger amounts of iron than contained in coal, varying with the tire type. All tires have steel wire in the beads, which is perhaps 3 to 4% of the tire weight. Steel-belted radial tires have substantially more wire in the belt—approximately 10% in all.

In the process of chopping a tire and converting it into tire derived fuel (TDF), some of the wire may be separated magnetically and removed. The fate of the tire wire during combustion varies, depending on the particular process used.

If whole tires are burned in a dedicated boiler such as the previously mentioned facility in Modesto, California (see Chapter 2), the steel melts, is collected as a slag, and is sold as scrap iron. When TDF is burned as a supplement to coal fuel in a conventional coal boiler, the steel also burns and contributes a significant amount of energy (3,500 BTU/lb). The iron oxide produced turns up in the ash. What occurs here is the reverse of the smelting process by which the iron was made from iron ore. The combustion of steel wire requires a high ignition temperature, which can be achieved in a coal boiler but not in a wood-burning boiler. When TDF is used as a wood supplement, the wire largely survives the combustion process and is collected on the grates. This can produce operational problems such as clogging of the grates; therefore, use of TDF as wood fuel supplement may face restrictions regarding the wire content.

One other significant compositional difference exists between coal fuel and TDF. Coal is impure carbon. Tire rubber is a mixture of carbon black with approximately twice as much hydrocarbon material. The sole product of carbon combustion is carbon dioxide. On the other hand, when we burn hydrocarbons, we generate one water molecule for each molecule of carbon dioxide produced. Because hydrocarbons are more calorific fuels (more energy per pound), it follows that for a given amount of energy produced, tires will emit significantly less carbon dioxide than coal. A.L. Eastman of the Goodyear Tire and Rubber Company has estimated this difference as 19.5%.[6-1] In the event of tight forthcoming regulations on carbon dioxide emissions, this could be significant.

Whole Tires as Boiler Fuels

Perhaps the first significant, practical process to burn whole tires as fuel was developed in Germany by Gummi Mayer, a major European retreader. In normal operation, a retreader purchases used casings from dealers. These casings have been inspected at the dealer's facility and are presumed to be retreadable. On reinspection by the retreader and especially after buffing off the remaining tread, the retreader will reject some of those casings as not being retreadable. The retreader has purchased some poor casings, and he must now dispose of them. Accordingly, it made sense for Gummi Mayer to design and build a small steam plant using the whole used casings as the sole fuel. High-pressure steam first was run through turbines to produce electricity and then was used to heat Mayer's presses. In this way, Mayer could produce all the steam he needed and half of his electrical requirements. When I visited Mayer's plant in 1977, it was in full operation and continues to this day. Combustion was complete, with result that the only product collected in the baghouse was zinc oxide of an off-white color and 90% pure. The steel wire melted, was collected as a metallic slag, and was sold as pig iron. Stack emissions were low and met German standards. Although modest in size, this plant was the engineering prototype for the 14.4 MW plant built for Oxford Energy by

General Electric and presently operating in Modesto, California, adjacent to the largest pile of scrap tires in the United States. (See Chapter 2.)

Steam boilers fueled with whole tires also have been built and operated by Goodyear in Jackson, Michigan, and in Wolverhampton, England. However, neither is presently in operation. Each facility was too small to serve as the primary facility in its area and was intended as a supplementary facility. However, even for this purpose, at that time no existing infrastructure could provide the requisite quantities of scrap casings at an acceptable price on an ongoing basis.

Boilers dedicated to using tires as the sole fuel, either as whole tires or TDF, do not appear to be particularly promising applications for many reasons. First, a high initial investment is required, compared to standard types of coal-fired boilers of standard design, construction, and output. Standard tire-fueled boilers do not exist and would have to be individually designed. Second, the limited availability of large quantities of tires at any given location imposes severe size restrictions on the plant size. As mentioned previously, the Oxford facility is a 14.4 MW unit. Major coal-fired power plants built today typically have a capacity of 500 MW.

On the other hand and as we shall soon see, whole tires are being burned in cement kilns in the United States in growing numbers.

TDF as Supplemental Coal Fuel

TDF is an excellent fuel supplement to coal in a stoker-fired boiler, using chips of a size comparable to the stoker coal fuel. The generating facilities of most large utilities burn powdered coal of 200 mesh size. In terms of cost, it is impractical to grind tire chips to that fineness, but stoker-fed units remain in common use and can use TDF in sizes from 0.5 to 2.0 in. In this size range, TDF can be produced at an acceptable cost. Stoker coal dollar costs can range from the low $20s to the mid-$50s per ton, depending on the location and the quantity required. Coal freight costs are high—in excess of $10/ton for short hauls—and can easily equal the cost at the mine portal for long hauls. I remember having to pay $55/ton for cheap Idaho coal delivered to Wisconsin by rail.

As previously noted, the sulfur level of tires is low. Because their heat of combustion is high, TDF can be used to upgrade fuel quality. Two practical problems occur when using TDF. First, TDF must be mixed with the coal in a fairly uniform way to provide uniform heat evolution and to prevent localized hot spots that could produce clinkers or damage grates. This increases costs in the coal yard. Second, old boilers that lack baghouses or other suitable means for particulate collection will release small amounts of zinc particulate and increase opacity in the stack gases.

Today, several utilities and manufacturers are generating steam using TDF as fuel or a fuel supplement, including a major new facility of the Illinois Power Company, which will burn TDF furnished by Waste Recovery Inc., from two new TDF facilities in Illinois.

TDF as Wood Supplement

Forest products industries commonly use their own wastes, such as bark, branches, and sawdust, as primary fuel. The heat content

Fuel Uses of Scrap Tires

of such wood is approximately 6,000 BTU/lb as normally used. This figure is seriously reduced after prolonged rainfall, but it can be upgraded with TDF as a supplement. This was the earliest practical volume use of tires as fuel in the United States. The practice dates back to a pioneering development by Mike Rouse and others at Boise Cascade in the Pacific Northwest and subsequently with Waste Recovery Inc. in Portland, Oregon, and in Dallas, Texas. The use of TDF has been widely accepted in the forest products industry and is well established there. The use is limited chiefly by the availability of steady supplies at competitive prices and suppliers. Because it being used as a supplement to a fuel with essentially negative cost, significant pressure is always placed on the price. Typically, it is sold for $20/ton or less.

TDF in the Portland Cement Industry

The manufacture of portland cement requires large amounts of energy. Approximately 500 lb of coal are required to produce 1 ton of portland cement. Accordingly, the cost of portland cement is heavily dependent on fuel costs and represents an attractive opportunity for TDF.

Two basic steps are involved in making portland cement. In the first step, limestone is heated, or calcined, to drive off carbon dioxide, leaving a smaller quantity of unslaked lime (calcium oxide). The weight loss is approximately 60%. In the second step, the calcined lime is mixed with sand, calcium sulfate, and small quantities of iron oxide and other ingredients. This mixture is heated to temperatures exceeding 2000°C (3632°F) for approximately 24 hours during passage down a long kiln. When finely ground, the fused product is portland cement. This ancient recipe and basic process have been virtually unchanged for more than a century, although some engineering process modifications have been made.

The preferred fuel is natural gas, which became too expensive during the petroleum shortage of the 1970s. Natural gas has been replaced with powdered coal, which is burned similarly to a gas with long flames extending into the horizontal kiln to heat the slowly moving stream of raw materials. Coal can be replaced by TDF in either step of the process but requires some engineering process modification in the calcining zone and in the preheater tower of the kiln.

It is interesting and fortuitous that two of the minor recipe ingredients for portland cement, iron oxide and sulfur, can be furnished by TDF. The steel is converted to iron oxide, and the sulfur dioxide cannot escape being scavenged by the hot lime and converted to calcium sulfate. From an environmental viewpoint, we have almost an ideal situation for tire disposal.

The use of tires as fuel for cement kilns was pioneered in Germany. The first major adoption was by Dyckerhoff Zement, who developed a process in the 1970s using medium-truck tires as the sole fuel. These were led up a long conveyor and carefully metered, through a complex system of baffles, into a 2000°K chamber. Precise control of the heat supply is necessary to control the reaction and to prevent fusion of the clinker. One problem associated with the burning of whole tires injected intermittently into the kiln was pulsation of gases as the tire vaporized. The decision to burn whole tires was made in default of any practical process to chop tires at that time. In winter, problems occurred with ice formation on the conveyor, and tires that had been stored outside were likely to contain an unwanted

Fuel Uses of Scrap Tires

lump of ice. By 1980 Dyckerhoff had three plants, burning 30,000 tons of scrap tires annually. In 1981, I toured the Wiesbaden plant where, for simplicity of operation, they burned only medium-truck tires, carefully weighed for close control of the fuel charge. The tires were transported via conveyor approximately 50 ft up the wall of the building and introduced into the kiln through a series of baffles to avoid undue loss of 2000°C (3632°F) heat. In winter, the problems were compounded by icing of the conveyor and by ice in the tire cavity.

Subsequently, other German cement firms adopted tires as fuel but used TDF with considerable process simplification. Use of TDF in cement kilns became and remains a major outlet for scrap tires in Germany. This widespread use of TDF rests partly on the fact that the cement kilns receive TDF free of charge. The costs of collection, chopping, and delivery of the product are borne by the tire industry.

Portland cement manufacture is a global business involving many large international producers. A number of the domestic companies manufacturing portland cement are the U.S. counterparts of European firms. In due course, they began to evaluate tires as fuel and are coming onstream in large numbers. This would appear to become the most popular fuel use. As of today (October 31, 1997), 36 kiln sites are burning tires; some kiln sites even operate with multiple units. Another dozen are conducting trials, and approximately 30 more are planning trials. Some are burning whole tires; others are burning TDF. The tires either are being supplied free of charge or the cement manufacturer collects a small stipend per tire. In this way, they are participating effectively in the tipping fee.

Reference

6-1 Serumgard, John R., and Eastman, Andrew L., "Scrap Tire Recycling: Regulatory and Market Development Pregress," presented at a meeting of the American Chemical Society, Washington, DC , August 23, 1994.

CHAPTER 7

Transportation Uses of Scrap Tires

Rubber in Asphalt

One of the most interesting and promising large-volume uses for rubber from scrap tires is as a road-building material, chiefly as an additive or supplement to asphalt. The use of asphalt (or bitumen) as a binder for stone and sand in road construction is attributed to John MacAdam (1756–1836).[7-1] Born in Scotland, MacAdam emigrated to the United States where he had a profitable business. MacAdam retired from his business, returned to England, and spent much of the remainder of his life building roads in England by the techniques that he had developed and that came to bear his name.

Originally, naturally occurring asphalts were used from large deposits such as Asphalt Lake in Trinidad. Today, commercial asphaltic materials are produced in huge quantities in oil refineries as by-products of distilling and cracking operations. These materials can be described as a complex mixture of condensed hydrocarbons, including one fraction (asphaltenes) that is essentially carbon. Asphalt is more or less soluble in other hydrocarbons and was used as an inexpensive filler or extender in rubber compounds many years before its application in roads.

The idea of using rubber as an asphalt modifier seems to have arisen in Europe. In 1938, the "Report of the Work of the Rubber Research Board" in England contained a report by M.W. Philpott on road surfacing trials using nitrite crumb rubber.[7-2] Dutch investigators Van der Bie and Wijnhamer published several papers on the application of rubber powder to asphalt road mixtures.[7-3, 7-4] In 1949, Dinsmore reported on an experimental stretch of roadway in Akron, Ohio, laid with a rubber-asphalt mix.[7-5]

During World War II, the Germans were using natural rubber (of which they had a plentiful supply) as a supplement for asphalt (which they lacked) in wartime road-building activities. On the American side, rubber was scarce, and we had to develop synthetic rubbers and practical procedures to produce it. We were able to do this with an enormous effort. Following World War II, as research and development on synthetic rubber began to decrease, many projects arose to incorporate rubber in roads. The earliest projects that I can remember were based on using natural and synthetic rubber latexes. These produced interesting results but also provided serious foaming problems and involved the use of costly materials compared to asphalt. The contemporaneous decline in the rubber reclaiming industry prompted a shift in emphasis, from reclaimed rubber to finely ground crumb rubber. Much of this development was pioneered by the U.S. Rubber Reclaiming Company, then in Buffalo, New York. Significant quantities of rubberized asphalt began to be used in tennis courts and running tracks. The successful commercial application of asphalt rubber compounds in highway construction arose from the work of Charles McDonald on Arizona highways, beginning in the 1960s. The work was protected by patents[7-6] which were widely licensed and have since lapsed.

To understand these developments, we should examine the compositions of asphalt, asphalt roads, and crumb rubber. Asphalt is a complex mixture of condensed hydrocarbons. Therefore, it is largely soluble in many organic solvents and is freely compatible with gum (unvulcanized) rubber over the entire range of possible compositions. As noted above, asphalt was used as an inexpensive filler and extender in rubber compounds for many years before any road activity involving rubber asphalt mixtures.

Of concern here are mixtures of asphalt—not with gum rubber but with crumb rubber, which consists of vulcanized rubber particles. The vulcanization process converts the more or less linear gum molecules into a three-dimensional network, and the vulcanized object acquires a permanent shape. It can no longer be dissolved by the same solvents in which it formerly was freely soluble. It can absorb substantial quantities of those solvents and be swollen by them, but it will remain particulate and retain its initial shape.

At room temperature, asphalts are soft plastic materials that become mobile liquids when heated to approximately 149°C (300°F). They are sticky and adhere strongly to the stone aggregate that constitutes the majority of the paving compound. The primary function of the asphalt is to hold that aggregate together in an integral structure, to fill the void space around the stone aggregate in order to resist moisture penetration, and to remain ductile over the normal range of temperatures to be encountered during the life of the road. Commercial asphalts with well-characterized sets of properties and in standard grades are available from several refineries. The choices to be made depend chiefly on the climatic conditions to be encountered. The asphalt binder should have enough viscosity to remain in place under high ambient temperatures, while not becoming embrittled at low temperatures.

Now let us consider how the addition of crumb rubber to the asphalt binder can affect its properties. If we mix a portion of fine-particle crumb rubber with warm asphalt, the first discernible effect will be an immediate rise in viscosity. As time passes, some of the asphalt constituent molecules will be absorbed into the crumb rubber, in the same way that the rubber would absorb gasoline, or fuel oil. This would enlarge the rubber particles and simultaneously increase the viscosity of the residual asphalt and consequently of the whole mixture. The extent of this process will depend on the particular composition of the asphalt, on the amount of crumb rubber particles and their size, and on the temperature and length of exposure. In time, the particles of crumb rubber could absorb their own weight from the asphalt chemical components and remain suspended in the thickened residual asphalt. Now, instead of a viscous fluid, a mat of rubber particles would develop, held together by the asphalt residues. This mat would have an entirely different set of physical properties than the asphalt fluid itself. For example, the fluid film can be squeezed thin by mechanical pressure and ruptured. Although the rubber particles may distort under pressure, they will remain particulate, will resist distortion, and will not allow themselves to be squeezed out.

With this picture in mind, we may consider the various uses of asphalt in paving compounds and the effect of crumb rubber modifiers on them.

Crumb Rubber in Crack Sealants

The first and simplest application of crumb rubber is as a crack and joint sealer. When cracks appear in the asphalt roadway, it is desirable to fill them to prevent the entrance of water. In concrete highways, we also require a sealant to be placed between adjacent paving blocks. Asphalt is a suitable material for this task. When hot, it can be poured or squeezed into the crack and soon sets to a strongly adherent semisolid. Asphalts come in a variety of viscosities, and some choices must be made. If the viscosity of the asphalt is too low, on hot summer days it may flow from the crack. On the other hand, if the asphalt has too high of a viscosity, it may freeze at low winter ambients and will then crack and come out. Adding 20 parts of crumb rubber, perhaps of 20 mesh size, will greatly raise the viscosity at elevated temperatures without adversely affecting the freezing point. In addition, the carbon black in the rubber provides improved aging properties. Crumb rubber modification of asphalts for crack sealants is widely practiced in most states, with large-volume use in the lower tier of states including Texas and California, and also New York and Pennsylvania.

Crumb Rubber in Repair Membranes

Another successful use of crumb rubber modification of asphalt involves the use of a membrane of asphalt rubber mix applied as a surface patch to repair a damaged, cracked highway surface area. After cleaning, the asphalt rubber compound is applied as a patch, covered with a fine stone aggregate-asphalt mix, and put into service. The flexible, elastic nature of the membrane can accommodate small amounts of additional crack growth and can prevent propagation of the cracks through the membrane to the surface. This technique was developed by MacDonald and his associates, and it was protected by patents which were licensed to others. The technology is called the Stress Absorption Membrane (SAM). The technique also is modified to permit deep repairs in damaged pavements. After removal of the damaged material and thorough cleaning, the asphalt rubber membrane is sprayed into place and then covered with several inches of a paving layer. Again, the elastic membrane, described as the Stress Modifying Membrane Interlayer (SAMI), is intended to discourage reflective cracking into the upper paving layer. These techniques have been widely used in some areas with great success. The utilization of crumb rubber involves more cost; however, in these applications, it apparently is cost effective.

Crumb Rubber in Paving Courses

The volume usage of crumb rubber in seals and SAMs is not trivial. However, it is diminutive when considered against the volume that would be required if crumb rubber were an important component of the main paving course. In a normal year, approximately 480 million tons of asphalt are used for paving materials. If crumb rubber were universally adopted at the moderate level of 3% (or 60 lb/ton of asphalt), 28 billion lb of crumb rubber would be required. Assuming that 10 lb of crumb rubber could be obtained from an average passenger tire, we would need more than 2 billion tires annually. This quantity greatly exceeds the total available supply. Clearly, the substantial use of crumb rubber in asphalt roadways could solve the problem of scrap tires.

Large-volume use of crumb rubber in asphalt paving requires that it be used in the main paving course. This is being practiced in Arizona and several other states where the use of crumb rubber is increasing steadily. Many states have been conducting trials, which may require observation over a period of several years before a judgment can be made.

What issues determine the value of crumb rubber in asphalt paving? One of the main issues revolves around the ability of crumb rubber to flatten the temperature-viscosity curve of the asphalt. Low-viscosity asphalts can have low brittle points. This is desirable in winter but undesirable in summer because the binder gets squeezed out, leading to rutting of the pavement. The addition of crumb rubber increases the viscosity of the asphalt. The stiffer binders also adhere better to the aggregate at elevated temperatures and resist rutting. Similarly, crumb rubber modification makes the binder less subject to flushing or bleeding to the surface. In addition, crumb rubber confers improved elasticity to the binder, which helps to prevent cracking of the pavement in cold weather. Finally, unmodified asphalt hardens on aging, partly because of oxidation. Crumb rubber modification improves the aging resistance of asphalt binder. This is attributed to the presence of carbon black in the rubber.

Opposed to these potential advantages are certain disadvantages in the use of crumb rubber modification. One major objection is increased cost. Crumb rubber is expensive (approximately $0.13/lb), and the use of crumb rubber usually requires a greater overall binder volume. Likewise, a considerable investment in equipment is required to ensure proper metering and mixing of the rubber. Moreover, the asphalt rubber blend is fussy, in that it usually requires a certain amount of time to reach a stable set of properties but cannot be stored indefinitely at elevated temperatures without severe deterioration. Finally, some of the better techniques were patented and are proprietary. These were widely licensed, but this adds royalty costs to the process.

In practice, several different processes are being used. These can be divided into two main groups—namely, wet process and dry process.

Wet process involves the metering and mixing of the crumb rubber modifier and asphalt in a reactor and allowing them to react for an appropriate period before the mix is added to the aggregate. This is the more common process and often is referred to as the McDonald process, although it was used by many earlier practitioners. Usually, a mixture of coarse and fine stone aggregate is used, carefully chosen to produce close packing and a dense mass. To this must be added exactly the right amount of hot asphalt to ensure that all the aggregate is coated with a film, but no more. Excessive amounts of asphalt are squeezed to the surface (bleeding), which is wasteful and produces an unsafe highway surface when wet. Too little asphalt will leave voids and provide inadequate coverage of the aggregate, resulting in water damage and stripping of the binder from the aggregate.

In the dry process, the crumb rubber, which is usually a wide mixture of particle sizes, is preblended briefly with the hot stone aggregate before admixture with the hot asphalt. Here, the intent is that the large rubber pieces, usually of 0.25-in. size, will replace and function as aggregate and not as binder. The effects can be profound. The presence of elastic members in the aggregate can reduce thermal and reflective cracking. A certain number of the coarse rubber particles are present on the highway surface, where they reduce tire noise, improve wet traction, and destabilize ice films for improved winter traction. This technology was developed in Sweden and brought to the United States under the name of Plusride. Certain designs and constructions are patented[7-7] and require licensing. Several states, especially California and Florida but also Alaska, have field installations undergoing evaluation. These installations seem to be highly successful to date. Other dry process systems have been developed and are undergoing study, notably the TAK system proposed by Barry Takallou.[7-8]

The wet process has a longer history and wider application than any dry process. Many miles of highway construction in several states attest to the potential value of crumb rubber modification of asphalt binders. However, not all projects have been successful; some have been conspicuous failures. Nonetheless, there has been a steadily growing interest in and acceptance of crumb rubber modified asphalt paving compositions, particularly in a few Southern states such as Texas, Florida, and California, and in Arizona, where it all began and where the use of crumb rubber modification has reached approximately 20% of highway construction.

ISTEA Mandate on Use of Crumb Rubber in Highway Construction

This picture was greatly altered in 1991 when Congress passed the Intermodal Surface Transportation Efficiency Act, commonly known as ISTEA (pronounced Ice Tea). Section 1038(d) of the Act mandated the use of asphalt rubber in federally funded highways, beginning in 1994 at the level of 5% and increasing to 20% of all federally funded highways in 1997. This action, in response to a growing national concern about the scrap tire problem, provoked a strong protest from many states that had no experience or insufficient or bad experience with asphalt rubber. An underlying objection was that underfunded state highway departments were being asked to accept a major cost increase and to help shoulder the cost burden of eliminating the scrap tire problem.

In response to these objections, Rep. Carr of Michigan introduced an amendment to the Appropriations Bill for 1994 forbidding the Federal Highway Administration from using its funds to implement, administer, or enforce the provisions of Section 1038(d) on the basis that it was an unproven technology. The 1995 Bill renewed the Carr Amendment. On this basis, it appeared that Congress was not going to require the early and perhaps premature adoption of rubber-extended asphalt but remains in a position to do so as the picture clarifies and as more information becomes available. What is clear is that rubber-modified asphalt is always going to be more expensive than unmodified asphalt and will be widely adopted only to the extent that it is demonstrated to be cost effective.

To help resolve this stalemate, in early 1995 the Federal Highway Administration let a million-dollar contract with the Engineering Research Department of the University of Oregon to conduct a definitive study of crumb rubber modified asphalt, including its technology and applications, its engineering soundness, and its cost effectiveness. Thirty-two states are participating in the study and the funding, which was commissioned under Section 1038 of ISTEA.

The study continues and is two-thirds complete. It will be presented to Congress as originally scheduled, notwithstanding that late in the 1995 Congress ISTEA was repealed, presumably as a matter of politics. An interim report will be made to the Federal Highway Administration in 1997. Two other recent developments of note that should be mentioned at this point are the following:

Transportation Uses of Scrap Tires

- The use of molded rubber-plastic blends into signposts and presumably later into fence posts (see Chapter 12).

- The use of cut-out sidewalls from medium-truck tires around and at the base of the bright orange plastic cones or barrels used to mark lane changes of roadway under repair. One nationally popular variety has a 20-in. diameter that nicely makes a cozy fit inside the tire sidewall which then rests on the cone flange and stabilizes it against tipping from minor impacts or misplacement due to high winds, a most welcome and serendipitous development that seems to be spreading like wildfire.

Rubber in Railroad Crossings

Another unique transportation-related use for rubber has been applied to railroad crossings.

At any intersection having a set of railroad tracks with a road or highway, we have the problem of finding a suitable paving material to occupy the space between and around the tracks. This paving material must permit smooth passage of a vehicle but cannot interfere with the safe passage of the flanged railroad wheels on the track. Many materials have been and are being used, including dirt and gravel, wood timbers, asphalt, and concrete. All of these materials have problems associated with them, and drivers have learned to cross tracks cautiously to avoid bumps and even a broken axle.

In recent years and in rapidly growing numbers, engineered rubber constructions are being used to pave railroad crossings in a way that provides a smooth, safe, and durable crossing for the vehicle. These rubber constructions are also satisfying the maintenance problems for the railroads, which periodically must remove the crossing materials to replace ties or rail, or to clean the stone ballast. The crossing materials must be firmly anchored in place to the ties and through them to the rail. Otherwise, the dynamics of the loaded wheels on the tracks produces a peristaltic pumping action, which would sweep unattached paving materials yards down the track.

Today there are three major manufacturers of rubber railroad crossings: Omni Products of Portland, Oregon; Kraiburg of America, a subsidiary of a German firm with a plant in Lisbon, Iowa, and corporate offices in Chicago, Illinois; and Goodyear Tire and Rubber Company of Akron, Ohio. Figure 7.1 shows a HiRail rubber railroad grade crossing manufactured by Kraiburg of America.

These three manufacturers use widely different engineering designs in fabricating and molding their products. However, the requirements for the rubber compounds are mainly to fill the space, and thus they can accommodate a great deal of tire scrap. In the past, the Omni product was made from tire buffings from tire retreaders, as were HiRail Crossings made by Kraiburg of America. However, the supply of tread buffings is limited. As crumb rubber becomes increasingly available at low cost, its use will materially increase.

Substantial volumes of rubber are required for railroad crossing applications. Full-depth rubber crossings, as supplied by Omni Products and Kraiburg of America, will weigh 275 to 350 lb/track ft. In 1994, rubber consumption in this application was close to

Transportation Uses of Scrap Tires

*Fig. 7.1. Rubber railroad grade crossing.
(Courtesy of Kraiburg of America)*

41 million lb. In 1996, industry sales were estimted at $12 MM and declining due to competition from precast concrete.

References

7-1 *Encyclopedia Brittanica*, 15th edition, 1990.

7-2 Philpott, M.W., "Report of the Work of the Rubber Research Board in 1938" (S.C.L., 18-11), pp. 25–26, 1938.

7-3 Van der Bie, G.J., and Wijnhamer, P.T., "Properties of Mixtures of Asphaltic Bitumen and Rubber for Roadway Purposes," *Arch Rubbercultuur*, 25:138-41, 1941.

7-4 Van der Bie, G.J., and Wijnhamer, P.T., "Application of Rubber Powder to Asphalt Road Mixtures," *Roads and Streets*, 84:36-40 (No. II), 1941.

7-5 Dinsmore, R.P., "Rubber-Asphalt Road Surfacing" (S.C.I. 27:763), *Canadian Chemical Process Industries*, 33:510-11, 1949.

7-6 U.S. Patent No. 3,891,585 to Charles H. McDonald, June 24, 1975; U.S. Patent No. 4,068,023 to Donald L. Nielsen and James R. Bagley, June 10, 1975; and U.S. Patent No. 4,085,078 to Charles H. McDonald, April 18, 1978.

7-7 U.S. Patent No. 4,086,291 to A. Natanael Svensson, April 25, 1978; U.S. Patent No. 4,548,962 to Gustaf Lindmark, October 22, 1985.

7-8 Takallou, M.B., and Takallou, H.B., "Benefits of Recycling Waste Tires in Rubber Asphalt Paving," Transportation Research Board Symposium, Washington, DC, January 1991.

CHAPTER *8*

Mats, Playturf, and Equestrian Uses of Scrap Tires

Rubber Mats

One of the most attractive and fastest-growing uses for crumb rubber from scrap tires is in rubber mats. Rubber mats are an old product. At one time, virtually all automobiles had black, rubber floor mats as standard equipment. During the 1950s, these rubber mats were largely replaced with more attractive, colorful, and cheaper vinyl mats.

The advent of substantial quantities of inexpensive crumb rubber and rubber buffings from retreaders has presented new opportunities. Large and growing quantities of rubber mats now are being manufactured in the United States and Canada for a variety of end products, many of which are new. Figure 8.1 shows examples of mats made from crumb rubber.

Techniques have been developed to combine large amounts of crumb rubber with small amounts of virgin rubber to produce high-quality rubber mats and other products. More surprisingly, it

Scrap Tires: Disposal and Reuse

*Fig. 8.1. Examples of mats made from crumb rubber.
(Courtesy of Royal Rubber and Manufacturing)*

now is possible to remold rubber buffings from retreaders and crumb rubber with small amounts of sulfur and other chemicals at elevated temperatures and an outer layer of virgin rubber compound to produce mats of merchantable quality for new markets. Of even more commercial importance is the fact that polyurethane compounds that cure at room temperature can be used to bond ground tire rubber into rubber mats having excellent physical properties without recourse to presses with heated platens.

Agrimats

One new development is the use of large rubber mats, known as agrimats, in cow barns and horse barns. These mats usually are 1 in. thick or more. The idea is to take advantage of the low thermal conductivity of rubber to provide a warm bed for the animals. The idea is hugely successful, and it is difficult to get the animals off the mats, day or night except at milking time and feeding time. In the case of cattle, there is economic value in terms of increased milk production. I am aware of some large dairies in which the cows are now milked three times per day with a 10% improvement in milk production. In a related development, pallets comprising a plastic envelope and filled with crumb rubber to a suitable thickness are being sold for the same purpose. In the Netherlands, according to a recent announcement,[8-1] Dunlop-Enerke will make waterbeds for cows. These are double-sided rubber mats with compartments to hold water. The cost of a cowmat in the United States is approximately $50. To justify this price, the mat must generate true value.

These mat-making techniques also are suitable for mats for horse trailers, bed liners for trucks, small mats for machine shops and factory operations where workers must stand for long periods, and many other small-volume items. This has developed into a multimillion-dollar business involving approximately a dozen U.S. and Canadian firms.

In addition to the growth in sales of conventional rubber mats, substantial commercial development has occurred regarding rubber mats in which crumb rubber particles are coated with a film of polyurethane rubber. The mixture then is squeezed into the desired shape and allowed to cure at room temperature and essentially at atmospheric pressure. Several different types of products can be made in this way. If the crumb rubber particles are small and uniform and the amount of polyurethane is adequate to fill the void space, we have a polyurethane molded object containing a rubber filler. If the crumb rubber particles are large and wet with only a thin film of polyurethane, we then obtain a porous structure of rubber particles bonded together with polyurethane. These structures are of particular interest for their energy absorption characteristics, and this has recently become an important concern with regard to children's playgrounds.

Rubber Mats for Playground Surfaces

Ten years ago, the Consumers Products Safety Commission took the position that asphalt, concrete, and packed earth were unsafe playground surfaces and should be replaced by wood chips, sand, or rubber matting. This recommendation had been largely ignored, in spite of almost 500,000 children under the age of 15 who were treated in hospital emergency rooms for playground accidents in 1995. In 1992, a Washington, DC, court trial produced a settlement of approximately $15 million in damages for a 10-year-old boy who fell 8 ft to an asphalt surface and who suffered brain damage as a result of his fall.

Mats, Playturf, and Equestrian Uses of Scrap Tires

The implications from this lawsuit have generated significant concern and considerable activity to reduce the liability of playground providers. ASTM Committee F-8 on Sports Equipment and Facilities has studied the matter and has established two ASTM procedures[8-2] to provide a specification and a test procedure to measure the "Impact Attenuation of Surface Systems Under and Around Playground Equipment." A peak deceleration of less than 200 G is specified, with an arbitrary parameter, the Impact Severity rating, of less than 1,000. If these specifications are not exceeded, a serious head injury resulting from a fall will be unlikely. The impact is a function of the height of the fall; the greater the height of the fall, the more cushioning will be needed.

Rubber-polyurethane mats can be designed to provide suitable safety levels for contemporary playground equipment. In practice, the usual approach is to cast in place around the equipment that is to be protected a polyurethane mat from coarse scrap rubber particles (e.g., 0.5-in. pieces) to a thickness appropriate to the height of the equipment (usually 3 to 6 in.). On top of this is placed a top coating (0.25 to 0.5 in. thickness) of a dense polyurethane surface coating, which may be colorfully patterned. These types of elegant mats are appearing in greatly increasing numbers in public playgrounds and in private facilities, such as McDonalds restaurants, which specify these types of surfaces where available.

In summary, the total potential sales volume for these playground mats is huge. During the 1995–1996 period, manufacturers expected to exceed $100 million in sales. No industry figures are kept, but the *1997 Scrap Tire and Rubber Users Directory* lists 33 firms that make mats from crumb rubber.

Playturf from Scrap Tires

As mentioned above, rapid growth has occurred in the use of scrap rubber mats for playground surfaces. In addition, loose rubber chips in the 0.5-in. size range make an excellent playground surface, better than sand or wood chips. This development has centered in the Midwest. Today, almost 100 playgrounds in Indiana and Ohio are covered with rubber chips. In a typical installation, the depth of chips is 3 in. or more, and usually 6 in. near climbing equipment where falls are likely to occur. The resiliency and energy absorption capabilities are the predominant virtues of rubber chips. However, they also are clean, nondusty, and nonstaining. This latter fortuitous effect stems from the age of the scrap tires from which the playturf is made; playturf made from new tires could produce staining. Figure 8.2 shows rubber chips being used as a protective surface on a playground.

The excellent consumer satisfaction and acceptance of rubber chips as a playground surface, where used, contrasts the slow rate of market penetration across the United States. Several concerns about playturf have interfered with its development, but these have been resolved.

First, some questions have been posed regarding potential leachate problems from scrap tire products stored in the ground. An early Minnesota leachate study on piles of scrap tires raised some questions and concluded that scrap tire products should not be used below the water level. Subsequent follow-up leachate studies alleviated these concerns. However, some states such as Michigan, by action of its Natural Resources Commission, did not endorse the use of rubber chips for playground surfaces or other outdoor use where leaching could be a problem. This

Mats, Playturf, and Equestrian Uses of Scrap Tires

Figure 8.2. Rubber chips as a protective playground surface in Denver, Colorado.

position discouraged sales in view of the implied threat that if you put down playturf, you might subsequently have to remove it. The ultimate authority in these matters is the U.S. Environmental Protection Agency (EPA) which, for a long time, remained inscrutable on this topic. EPA had refused to declare scrap tires a hazardous waste in spite of frequent urging to do so, but took no positive action until early 1994 when it published a document essentially endorsing crumb rubber for uses such as playturf and soil amendment.[8-3] In view of this action, we might expect more widespread use of playturf during the next few seasons.

A second concern about the widespread adoption of playturf is that it is flammable. A playground is not spontaneously combustible, but it can be ignited with a match. Several cases of this kind of vandalism have been reported. Once detected, a playground fire should not be a great problem. Playturf burns with a flame height of only 4 to 5 in. and advances slowly—only several feet in one hour. The flame is easily extinguished with water from a hose, and the fire can be easily confined by raking away the playturf to leave a gap in front of the slowly advancing flame front. An untended playturf fire could ignite and burn the playground equipment, as occurred in a playground in the Chicago area. In the resulting litigation, the court took the position that as long as the flammability of the playturf was featured in the product label, a product liability problem did not exist.

For several years, the Used Tire Recovery Program of the Illinois Department of Energy and Natural Resources advocated the use of wire-free tire chips as a loose-fill playground surface. They commissioned TRACE Laboratories to conduct a study of the impact attenuation properties of rubber chips, compared to other common loose-fill playground materials. The report, published in January 1994,[8-4] shows that at a depth of 6 in., rubber chips were approximately twice as good as the next best material (wood chips) using the ASTM test procedures previously cited.[8-2]

In the 1995 season, sales of playturf exceed 1 million lb. A sharp increase was expected for the 1996 season, but no industry figures are kept. However, there are many more suppliers listed in the *1997 Scrap Tire and Rubber Users Directory.*

Rubber in Equestrian Arenas

Rubber chips also are beginning to find a place in equestrian arenas, for reasons similar to their use in playturf. (See Figure 8.3.) Equestrian arenas, especially indoor ones where sand is frequently

Figure 8.3. Rubber chips being used in an equestrian arena.

used, are dusty and normally require an indoor sprinkler system to control dust. Approximately 75 years ago, spent hemlock bark from hide tanning operations was spread on these arenas to control dust. These arenas were known as tan bark arenas. However, this practice ended with the disuse of hemlock bark for tanning. Today, a substitute product from bonded wood particles has found some use here, but it is expensive and can deteriorate. A growing number of arenas are finding that a 3- to 6-in. layer of rubber chips up to 0.5-in. size provides a satisfactory surface layer. Because the rubber chips are moisture resistant and nonwicking, the dust problem is substantially alleviated. If a rider is thrown from a horse, the fall is softer and the likelihood of injury is less. For this use and for playturf, the rubber chips must be rigorously free of wire. However, in equestrian arenas, the horses experience better footing if the rubber chips contain some residual fabric imbedded in them from the tire carcasses.

A recent patent, issued to Robert Malmgren[8-5] of Fort Collins, Colorado, demonstrates additional values if the rubber chips are rototilled into the soil of the arena. This development is becoming popular in Colorado and the adjacent areas.

One of the major obstacles in using rubber chips in equestrian arenas has been the high cost. To equip most arenas, the price ranges from $5,000 to $10,000. However, as the price of rubber chips decreases, the cost picture should improve.

References

8-1 "Dunlop-Enerka Waterbeds Making Cattle Comfortable," *Rubber and Plastics News II*, p. 4, May 12, 1997.

8-2 ASTM F 355-86 and F 1292-93, ASTM, 1916 Race St., Philadelphia, PA 19103.

8-3 U.S. Environmental Protection Agency, Part V, 40CFR Part 247, Comprehensive Guideline for Procurement of Products Containing Recovered Materials, May 1, 1995; Part VI Recovered Materials Advisory Notice, May 1, 1995.

8-4 "Impact Attenuation Test of Wire-Free Tire Chips Used as a Loose-Fill Playground Surface," TRACE Laboratories, 4631 N. Olcott, Chicago, IL 60656, January 1994.

8-5 U.S. Patent No. 5,020,936 to Robert C. Malmgren and Edward L. Umlauf, June 4, 1991.

CHAPTER 9

Scrap Tires in Sewage Sludge Composting and Soil Amendments

Sewage Treatment Plant Sludge

Conventional municipal sewage treatment plants purify human and other domestic wastes. These treatment plants work in this way: First, any filterable metal, plastic, glass, and other inorganic objects are removed from the raw sewage as received. Next, this filtered sewage is confined to a tank in the absence of air (anaerobic), and fermentation occurs there over several days. The sewage is then transferred to another tank for additional fermentation in the presence of air (aerobic). Following this, it is allowed to settle. The large volume of water, now purified and potable, can be discharged. The thick, mucilaginous waste product that has settled out of the tanks is known as sludge and requires final treatment.

Often sludge has been discarded in open fields, but this is objectionable from an environmental viewpoint because it is malodorous, contains weed seeds, and composts slowly. Preferentially, the sludge is burned. However, because of the high water content (75%), burning the sludge requires large amounts of oil fuel and therefore is expensive. Likewise, if not done properly, burning sludge can create a stench in the neighborhood.

Around 1975, the U.S. Department of Agriculture at its Beltsville, Maryland, laboratories under the leadership of Elliot Epstein developed an elegant process to compost sewage sludge. This process consisted of mixing the nitrogen-rich sewage sludge with a carbonaceous material (usually wood chips) as a bulking agent and then percolating air through the mixture. The wood chips serve a dual function. First, as a bulking agent, their platelike shape forces the percolating air into dispersed, reticulated pathways through the sludge, ensuring effective penetration of oxygen throughout the entire mass of sludge. Any portions of the sludge that are not aerated will undergo anaerobic fermentation and consequently will produce offensive odors. Second, the basic carbohydrate nature of the wood chips provides energy for the fermentation process. A vigorous fermentation reaction ensues, generating a large amount of heat and incidentally driving off much of the water. When the reaction is finished (usually after several weeks) and the mass has cooled, what remains is a friable mass that is relatively dry (25 to 35% moisture). This mass may be sifted to recover what remains of the wood chips for reuse. The final product is high-grade compost suitable for horticultural use.

Tire Chips Replace Wood Chips in Composting

During the 1970s, this composting process was considerably developed in one of two municipal sewage plants located in Windsor, Ontario.[9-1] (See Figures 9.1 and 9.2.) In initial

Sewage Sludge Composting and Soil Amendments

Fig. 9.1. Aerial view of sewage sludge composting facility at the West Windsor Water Pollution Control Plant in Windsor, Ontario, Canada. (Courtesy of L.S. Romano, Pollution Control Corporation, City of Windsor)

Fig. 9.2. Close-up of a windrow at a sewage sludge composting facility at the West Windsor Water Pollution Control Plant in Windsor, Ontario, Canada. (Courtesy of L.S. Romano, Pollution Control Corporation, City of Windsor)

operation, the sewage sludge, at approximately 25% solids, was mixed with three volumes of wood chips and piled on top of perforated plastic pipe in long windrows to a height of approximately 12 ft. For a few days, air was sucked through the mass into the pipes to contain any odors. Then air was blown through the mass for several weeks. The fermenting mass soon reached temperatures of 65°C (150°F) and higher. The effluent air carried off much moisture, and the whole windrow subsided to a height of approximately 10 ft. When the composting reaction was spent, the product was run through a gravel sifter to recover what remained of the wood chips, which also were subject to composting and normally lasted approximately three cycles. The compost ideally was a friable material containing approximately 35% moisture but sometimes required additional drying in heated bins.

To economize the use of expensive wood chips, Windsor began to replace some of the wood chips with 2- to 3-in. tire chips. Over a year, the amount of rubber chips was progressively increased, ultimately replacing 50% of the them. A substantial savings of wood chips was achieved. During one year, the savings effectively paid for the tire chips, which could be reused indefinitely. The chief disadvantage associated with the tire chips was a relatively high level of iron and zinc in the compost, arising from the dissolution of projecting short lengths of wire from the steel belts and beads of the tires and the extraction of zinc from the rubber compound. The high iron and zinc levels declined sharply after the first cycle of use and reached normal levels after approximately three cycles.

In addition to the considerable cost savings, the rubber chips were of significant benefit to the process. We have noted that, as fermentation proceeded, the height of the composting heap decreased as the water evaporated and the evolved gases escaped. However, when substantial amounts of rubber chips were present, the subsidence of the pile was markedly reduced—from several feet to less than 1 ft. The tire chips did not compact as the wood chips did, but they produced a looser internal structure with more porosity. This facilitated the escape of the water and consequently produced a dry compost product not usually requiring further drying.

A second benefit of the rubber chips resulted from the low thermal conductivity of rubber. The pile retained heat better and remained warmer in winter. This allowed the composting reaction to proceed through protracted cold spells, when composting essentially would have stopped if only wood chips had been used as the bulking agent.

Several hundred municipal sewage treatment plants now compost sewage sludge. However, to date, the adoption of tire chips as a supplementary bulking agent has been slow, in spite of favorable reports from extensive research conducted at Rutgers University by Andrew J. Higgins and colleagues.[9-2] However, some substantial composting operations use rubber chips, with favorable results.

In August 1995, a new sewage treatment facility in Davenport, Iowa, began using rubber chips in its composting operations. The results were favorable, including considerably reduced operating costs. I was told that each successive time the tire chips were reused, a savings of $10 worth of wood chips resulted. Another long-established composting facility in Columbus, Ohio, began to purchase rubber chips in the same year and presumably is using them in the same way. Also, one commercial composting operation, A and M Composting in Manheim, Pennsylvania, is a major user of tire chips almost to the exclusion of wood chips.

Sewage sludge composting is not expected to become a major outlet for tire chips simply because only one purchase is required and they can be reused indefinitely. The initial purchase is large, however. To convert the currently operating sludge composting plants from wood chips to an appropriate usage of tire chips would require one year's total supply of scrap passenger tires. The savings produced would be enormous because each time a ton of tire chips is reused, it would replace another ton of wood chips, at a savings of approximately $10 per ton.

Effects of Crumb Rubber on Soil

Several state universities—among them Missouri, Iowa, Arkansas, and Mississippi State[9-3]—have had ongoing programs to determine the effects of crumb rubber, buried in the soil, on a variety of plants. Negative effects sometimes were encountered, especially in soils having a pH lower than 5.3 This was attributed to the effects of high zinc concentrations because significant amounts of zinc can be extracted from the crumb rubber at low soil pH. No important benefits were observed on any of the crops investigated.

In the 1980s, Malmgren, Saltanpour, and Cipra began to study the effects on turfgrasses of crumb rubber that had been rototilled into the soil together with quantities of suitable compost. They observed that root penetration was faster and that water penetration of the soil was enhanced. Most importantly, they noticed that soil treated in this way resisted compaction from surface traffic. They realized the practical implications of this observation and obtained a U.S. patent.[9-4]

To understand this benefit, we should explain the problems created by surface traffic. A steady pattern of foot traffic in a narrow zone across a field of grass soon produces a footpath, where the grass no longer grows. The same familiar phenomenon also occurs with vehicular traffic. The explanation rests partly on the physical damage done directly to the grass plants, but more importantly also to the progressive compaction of the soil in which the plants grew and the consequent exclusion of air. Plant roots need air as much as they need water. This explains why most plants will not grow in flooded areas and why trees are killed by protracted flooding.

Crumb Rubber as a Soil Amendment

After a hiatus of more than a year, the commercial development of crumb rubber as a soil amendment using the Malmgren patent[9-4] was undertaken by JaiTire Industries, Inc., a startup company located in Denver, Colorado. JaiTire called its product Rebound. The product consisted of a mixture of approximately equal parts of commercial compost and crumb rubber particles smaller than 0.25 in. down to 10 mesh. The precise composition of the Rebound to be used was determined after an appraisal of the soil conditions at the site. After three years, in early 1996 American Tire Recyclers acquired control of the patent and the Rebound logo and is continuing the commercial development of Rebound on a national basis. Rebound is not necessarily a panacea—some soils present special problems, and preliminary appraisal by a soil specialist is necessary to prevent some failures. Often, however, the results are spectacularly successful.

Scrap Tires: Disposal and Reuse

One project[9-5] conducted in 1994 in Lancaster, California, involved construction of two softball fields in the Lancaster City Park. The fields, which are adjacent to each other, are identical in all respects except that one field contains Rebound in the outfield and the other does not. Thus, the two fields provided an ideal test site. Over a five-month period, the results were monitored by Professor Kurtz, a professor of Turfgrass Science at California State Polytechnic University at Pomona.

In the field containing Rebound, the perennial ryegrass seed germinated a week earlier than the seed in the field that did not contain Rebound. This occurred presumably because the soil temperatures in the field containing Rebound, which were measured at two depths, were consistently higher than those in the field that did not contain Rebound. The differences persisted until a full cover of grass was available to screen the rubber particles from the sun's rays. Root penetration also was faster in the field containing Rebound, and roots reached greater depths than in the companion field.

Likewise, water infiltration measurements showed that the field containing Rebound was more permeable and drained more quickly than the field without Rebound. Moreover, the differences increased during the five-month period, perhaps attributable to the greater root-depth penetration of the field containing Rebound. This feature is particularly meritorious in outdoor athletic fields.

Impact absorption and soil hardness tests also showed that the field containing Rebound was consistently softer at several different moisture levels. These numbers indicate a safer playing field.

All types of playing fields are candidates for using rubber chips as a soil amendment. Football fields, baseball infields, and soccer fields are particularly suitable because of the concentration of traffic in the areas in front of the goals. On golf courses, the areas frequented by golf carts, where there is no cart path, can be greatly improved by the addition of rubber chips, including the areas at the end of the asphalt cart paths where the grass does not grow. Similarly, public parks with a high volume of foot traffic are responsive to the addition of rubber chips, particularly because many of these parks require frequent resodding. For example, a single rock concert may totally destroy a lawn, necessitating immediate resodding. Figures 9.3 and 9.4 show crumb rubber being applied as a soil amendment to improve grassy areas.

Several other gratuitous advantages justify the use of rubber chips as a soil amendment. Less need for watering is frequently reported, and water penetration is greatly increased in fields with poor drainage. Fields that often are unusable after heavy rainfalls drain much more rapidly when rubber chips have been used as a soil amendment. A striking example is the Jacksonville, Florida, Metropolitan Park Amphitheater, in which Rebound had been installed by American Tire Recyclers, Inc. Despite earlier torrential rains, the 1994 Jacksonville Annual Jazz Festival was able to proceed on schedule on a field that otherwise would have been unusable except for the installation of Rebound.[9-6]

The Turfgrass Department at Michigan State University, under the leadership of Professor John Rogers III (Trey Rogers), has been very active in the broad investigation of crumb rubber as an aid in turfgrass cultivation. For example, they were able to show that a well-worn student path across a campus lawn could be restored by coring it with a conventional coring machine and

Fig. 9.3. Crumb rubber top dressing being applied at trouble spots at the Denver Country Club in Denver, Colorado.

*Fig. 9.4. Application of crumb rubber on a practice driving tee at a golf course in southeastern Michigan.
(Courtesy of David Kenyon)*

then sweeping crumb rubber into the holes after the earthen cores were removed from the field. The porous structure of the rubber clippings provided a fairly permanent means of aerating the subsoil.

Some of the definitive studies in the development of Rebound were also conducted on Michigan State's turfgrass farm in East Lansing, Michigan. More recently, their interest has been concentrated in the utilization of crumb rubber as a top dressing for turfgrasses. Crumb rubber is not easily wet by water nor does it absorb water. Thus, crumb rubber can serve as an ideal mulching material and top dressing. Various particle sizes and depths of dressing were studied. Three successive layers of rubber chips 0.25 in. or smaller were particularly successful in protecting the growing crown of the plant with a 0.75 in. overall thickness rubber wall. Protection is not perfect; at high-traffic levels, the turf will suffer. However, with normal watering, the turf shows quick recovery to normalcy. On the other hand, high traffic on untreated areas usually requires complete resodding (e.g., after a major golf tournament). These high-traffic areas are not necessarily areas of play on the course but are locations where the gallery traverses the fairway or around the refreshment stands or toilet areas.

Trey Rogers and his colleague were recently awarded a U.S. patent for this development.[9-7] In anticipation of this turn of events, JaiTire Industries, Inc. has licensed the patent from Michigan State University and is attempting to sell it through a nationwide network of dealers. The early results are quite favorable. The idea works, and demonstration is easy. Application also is easy and does not require tearing up the golf course as was required with Rebound. Also, testing programs can be tailored to budgets. You can try a few tons in a problem area, and when it works, you can proceed to using truckload quantities. The commercial product is trademarked "Crown III." At present, it is part of the program on more than 500 golf courses in the United States and others in Canada and abroad. Crown III may become one of the major uses for scrap tires in the next few years. It is expected that a minimum of 6,000 tons will be sold in the 1997 season, which is more crumb rubber than was sold for all purposes in 1996.

Crown III was used with success on the Oakland Hills Golf Course in Birmingham, Michigan, at the time of the 1997 National Open. Pat Knox, who applied several tons the day before the tournament opened, said that as a result of an all-night rain, he expected to find it all washed out onto the adjacent Maple Road the next morning. However, the 1.13 specific gravity of crumb rubber sufficed to prevent it from floating away. Crown III was also used successfully on the course at the time of the Ryder Cup international matches one year earlier. Its virtues are not confined to golf courses. A single application of Crown III to the worst municipal football field in Tulsa, Oklahoma, converted it in one season to being the best field. Another installation is planned for the current year. It is now being evaluated in an athletic field in the U.S. Military Academy at West Point in New York, with a view to extending it to the Parade Ground next year if all goes well.

References

9-1 "Rubber Tire Blend Works in Windsor," *Biocycle*, Vol. 24, No.6, pp. 44–46, 1983.

9-2 Andrew J. Higgins *et al.*, "Shredded Rubber Tires as a Bulking Agent for Composting Sewage Sludge," National Technical Information Service, 525 Port Royal Rd., Springfield, VA 22161.

9-3 Jensen, R.D., Glenn, R.C., and Ward, C.Y., "The Effects of Admixtures of Ground Rubber on Some Physical and Chemical Properties of Soil Materials and Plant Growth," unpublished research report, Mississippi State University, 1971.

9-4 U.S. Patent No. 5,014,462 to R.C. Malmgren, May 14, 1991.

9-5 Proceedings of Scrap Tire '94: Scrap Tire Business Development Conference, September 22–24, 1994, Tacoma, WA; Scrap Tire Management Council, 1400 K St. N.W., Washington, DC 20005.

9-6 *Scrap Tire News*, Vol. 9, No. 1, p. 1, 1995.

9-7 U.S. Patent No. 5,622,002 to J.N. Rogers III and J.T. Vanini, April 22, 1997.

CHAPTER *10*

Civil Engineering Studies and Applications

Chapter 5 identified some novel properties of tire chips. Among these are some of special interest in civil engineering applications, particularly the low density and durability, which suggest the use of tire chips as lightweight aggregate and fill material. This chapter will elaborate on a few of the applications recently investigated along these lines. Professor Dana Humphrey of the University of Maine and Professor Tuncil B. Eder of the University of Wisconsin, among others, have made careful studies and have reported on the specific parameters of tire chips, including the density, compressibility, compressed density, shear strength, friction angle, and permeability.[10-1,10-2]

Use as Lightweight Aggregate in Fill

Some of the first work regarding the use of tire chips as aggregate was done in Minnesota, where roads had to be constructed over peat bogs. Use of conventional fill materials, with a density of approximately 2.5, would result in a road that sunk below grade with the passage of time. Use of tire chips as fill, with a density of approximately 1.15, solved that problem but raised some questions about leaching. This inspired a frequently referenced leachate study,[10-3] somewhat flawed in my opinion because of its reliance on old tires recovered from old tire piles where they appear to have been contaminated with heavy metals not normally used in tire manufacture. The study concluded that tire chips were acceptable fill but should not be used below the level of the water table.

A more common problem arises when a road is to be built on a hillside where the terrain has a lateral slope and the application of quantities of fill is required to achieve a level road surface. Depending on the degree of the slope and the stability of the terrain, the possibility of a landslide exists. This possibility can be considerably reduced through the use of low-density aggregate.

In 1990, the Oregon Department of Transportation conducted an experimental project using shredded tires to correct a landslide under a highway embankment.[10-4] Approximately 8,500 yd^3 of earth embankment, weighing 12,800 tons, was replaced with a similar volume of shredded tires weighing 5,800 tons. The embankment has a 3-ft compacted soil cap on the top and slope and also supports a conventional aggregate base and asphalt pavement.

In l993, Maine began an evaluation of tire chips as fill beneath a paved road[10-5] in North Yarmouth. The evaluation had two objectives. One objective was to determine the minimum thickness of soil fill that is needed between the top of the tire chips and the bottom of the pavement to limit pavement deflection to acceptable levels. The other objective was to make a leachate determination of the effects of tire chips on the water quality when tire chips are placed above the existing groundwater table. Preliminary indications from this ongoing study are that pavement deflection is not a problem and that leachate has not exceeded applicable standards for water purity. Figure 10.1 shows tire chips being used along S.R. 9 near North Yarmouth.

Fig. 10.1. Tire chips used along S.R. 9, Township 31 MD, Maine. (a) Unloading the 3-in. tire chips; (b) Compacting the tire chips with a smooth drum vibratory roller; (c) Covering the completed tire chip layer with a nonwoven geotextile layer. (Courtesy of Dana N. Humphrey, University of Maine)

Use as Subgrade Thermal Insulation

In 1991, under the leadership of Professor Humphrey, Maine began an investigation of layers of tire chips as subfill insulation beneath a gravel-surfaced road.[10-5] Two different thickness layers of tire chips were used: 152 mm and 305 mm (6 in. and 12 in., respectively). There also were three different levels of granular soil overlay: 305 mm, 457 mm, and 610 mm (12 in., 18 in., and 24 in., respectively). In each of two winters (1992–1993 and 1993–1994), all test sections showed major reductions in the depth of frost penetration compared to the adjacent control sections. More importantly, they showed substantial reductions in the extent of frost heaving during the thaw period, amounting to almost a fivefold reduction in one test section. From these results, we can conclude that the use of tire chips as subgrade insulation offers excellent potential as a cost-effective way to improve the trafficability of gravel-surfaced roads in northern climates during the spring thaw.

Use As Backfill Behind Retaining Walls

Professor Humphrey noted the potential utility of tire chips as retaining wall backfill.[10-1] Their low compressed density should result in lower horizontal pressures and thus permit thinner retaining walls of simpler construction. The Colorado Department of Transportation reported favorably on one such retaining wall[10-6] that was backfilled with tire chips but which subsequently failed (*vide infra*).

Use in Landfills

Perhaps the most widespread and fastest-growing engineering use of tire chips is in landfills. Current environmental regulations in the United States require landfills to be lined with clay walls or plastic film to protect neighboring groundwater sources from undesirable leachate constituents. These in turn require protection against rupture damage, and geotextiles are used as supplementary protection. Increasingly, landfills also are lined with several inches of tire chips to protect the geotextile layer and to provide good drainage in the landfill. Rubber chips also are used as landfill cover and on the sloping sides of landfills. Under normal compression loads from earthen fill, the tire chips have excellent permeability (similar to that of clean gravel); on slopes, they remain in place better than sand, presumably because of the friction properties of rubber chips. Waste Management, Inc. and Browning Ferris both use large amounts of tire chips in their landfills. A single purchase order may call for 30,000 tons of tire chips, the equivalent of more than three million passenger tires.

Use in Septic Fields

A 1990 study in Vermont[10-7] demonstrated that tire chips can be an effective replacement for crushed stone in septic field drainage systems. Crushed stone is viewed as a limited resource. With tire chips, permeability was satisfactory, and the storage capacity of a drainage field of tire chips would be greater than one with crushed stone. Evaluation of the available leachate studies indicated no problem with leachate. In addition, it was noted that the lighter weight of tire chips over stone would be an advantage in regard to transportation costs. A geotextile cover on both the stone and tire chips was desirable to prevent intrusion of the overlaying soil.

Also in 1990, the Iowa Department of Natural Resources commissioned a study on the use of shredded tires as a replacement for drainage soil in leachate collection systems at municipal landfills.[10-8] Hydraulic studies showed a safety factor of at least 800 in the permeability of tire chips compressed under a load equivalent to a 35-ft head of landfill waste.

Recent Problems in Using Tire Chips in Highway Construction

During the past decade, more than 70 apparently successful highway projects have used tire chips in substantial quantities to replace more conventional, heavier earthen materials. As a result of this success, the scrap tire people have pinned great hopes on civil engineering applications as a major element in the total solution of the scrap tire problem, and many new projects were planned. However, three projects developed hot spots because of exothermic reactions in the rubber mass, which culminated in fires. These constructions had to be torn up to disperse and cool the rubber masses and extinguish the flames. The projects then had to be repaired with more conventional materials, at considerable expense.

Brief descriptions of these three failed projects are as follows:

1. In 1994 in Glenwood Canyon, Colorado, the Department of Transportation built a 70-ft-high retaining wall from rubber blocks manufactured from scrap tires and used tire chips as backfill. In summer 1995, shortly after completion of construction, steam began to issue from vents. In October, the wall caught fire and had to be torn down. No satisfactory explanation of the cause of the fire has been found.

Civil Engineering Studies and Applications

2. In Falling Springs Road, Garfield County, Washington, a hairpin turn was located around the head of a ravine. Over a culvert placed on the ravine floor, a 50-ft embankment of tire chips was made, and the road was constructed over it. The tire chips used in this construction came from two different sources and were somewhat different in character. The first portion was produced by a hammermill, and the gross pieces were difficult to distribute evenly by the bulldozer. The remainder, including much of the upper portion of the embankment, was made from chips produced by a conventional machine that sheared the tires. Construction began in fall 1994 and was completed in spring 1995. In October, steam began to be emitted. Open flames were observed sporadically in January and continued into February 1996, when the project finally was torn apart.

3. In S.R. 100 Loop Road in Ilwaco, Washington, a landslide occurred in December 1994, leaving a 140-ft gap that was 25 ft deep. Tire shreds were determined to be the most cost-effective way to fill this gap because their light weight eliminated the need for a shear key to be cut into the underlying soil or rock, or the need for construction of a buttress at the toe of the fill. Construction began in September 1995. Approximately 4,000 tons of tire shreds were placed over a two-week period in October, and the road was paved at the end of October. By the end of December, a crack in the pavement was observed. On January 3, 1996, steam was observed and heat could be felt on the road. A limited amount of temperature, air quality, and settlement monitoring began on January 17. In February, the reaction rate was observed to be increasing. In March, liquid pyrolysis products began to

issue from the base of the fill, leading to a decision to tear down the project.

In the face of these unfortunate events, all similar projects were put on hold. The Federal Highway Administration announced that it would not approve any new projects until the problems were understood and resolved.

As a first step, John Serumgard, Chairman of the Scrap Tire Management Council, organized an *ad hoc* committee to study the matter. This committee held several meetings. Professor Dana Humphrey of the University of Maine visited all three sites, [10-9] making a retrospective examination of potential causes, and prepared a report for the Federal Highway Administration. However, the remains of the fires provided no clear key to the origins. Tire piles are not normally subject to spontaneous combustion nor are piles of tire chips. However, a number of people who chip tires have encountered exothermic reactions in piles of freshly cut chips and have generally associated them with prolonged contact between hot portions of wire and fine particles of overheated rubber. The remedy is to disperse and cool the site; this seems to eliminate the development. By hindsight, it can be surmised that at least one of these projects did have a quantity of 'fines' introduced that could have become hot. It also appears that some organic materials or fertilizers also could have been imbibed by the tire pile. Bacteriologic oxidations also come into question here because very large masses of tire chips were involved and made a splendid blanket to insulate even a small source of heat while it grew into a conflagration. Because piles of tires or piles of tire chips are not generally subject to spontaneous combustion, our problem is to identify some chemical or biological event capable of raising the temperature of the mass to the ignition temperature of rubber, which is approximately 500°C (932°F).

Here we may have been assisted by a fourth tire engineering project, also in the Northwest, which became hot after completion. In Eugene, Oregon, Spencer's Creek Drive provides the only access to 15 residences in the Spencer's Creek Subdivision. It was built in 1979 using fill of marginal shear strength and experienced a shear slope failure in January 1995 after a period of record precipitation. After study, it was decided to use rubber chips as lightweight aggregate in the reconstruction, which was completed before the end of 1995, using approximately 600 tons of tire chips provided by Waste Recovery.

The planned inclusion of temperature monitoring equipment was delayed until April 1996, when it was discovered that heating had begun and the temperature was 42°C (108°F). The temperature continued to rise over the summer to slightly lower than 50°C (122°F) but began to decline as winter approached. In 1997, the temperatures began to rise again as the weather grew warmer; however, the temperatures did not rise as rapidly as the preceding year. Dr. Gunnar Schlieder of Gem Consulting Inc.,[10-10] who monitored the events, observed the disappearance of the wire from the tire chips and was able to extricate from the hot zone via the aperture to the temperature sensing zone. Using a magnet, he also retrieved a quantity of what we take to be black iron oxides, mixed with small pieces of wire, which will be further examined. In his opinion, the quantity of wire remaining in the chips was not sufficient to raise the temperature to the same level as the previous year. This will allow the opportunity for adequate study of the heating reaction as it subsides and sampling of the gases in the cavity also.

Scrap Tires: Disposal and Reuse

The *ad hoc* committee held a meeting with personnel from the U.S. Army Corps of Engineers, which expressed interest in the project and a willingness to participate in the investigation if funding becomes available. The committee also is preparing a set of temporary specifications of a conservative nature to allow work to resume at least on a limited basis in the near future. A main precaution will be to limit the thickness of the body of chips. Another will be to specify large chip sizes (2 and 4 in.) to minimize the amount of exposed wire. Fine rubber particles will be excluded, as well as fertilizers and organic materials. This tentative temporary specification will be presented to the Federal Highway Administration for its review and approval.

In these uses of tire chips in highway construction as subgrade fill or as backfill behind retaining walls, the primary concern is not with the fire that may occur at the end of the process, but with the nature of the reaction that allows a heat buildup from ambient temperatures to the ignition temperature of the rubber. Large tire piles or piles of tire chips generally are not thought to be subject to spontaneous combustion. Nonetheless, in the confinement of the landfill, some considerable oxidation processes are occurring. Possible explanations are as follow:

1. Oxidation of the steel to iron oxides may be occurring. In aged piles of tire chips, the exposed steel wire is known to rust off, in a moderately exothermic reaction. This could be subject to catalysis by water and other metals.

2. Composting of organic materials that may have been included incidentally with the fill may raise the temperature to the point where oxidation of the rubber can begin.

3. Placement of rubber fines (airborne dust) from the tire chopping operation into the landfill together with freshly ground metal chips has been known to produce fires in the tire chopping plant.

4. Biological oxidation of the rubber may be caused by microbes, yeast, fungi, or other microbes present in topsoil in the landfill, assisted by leachate of fertilizer materials.

A major factor in the process is the low thermal conductivity of the rubber chips, which prevents the escape of heat from the tire pile.

References

10-1 Humphrey, D.K., Sandford, T.C., Cribbs, M.C., and Manion, W.P., "Shear Strength and Compressibility of Tire Chips as Lightweight Backfill for Retaining Walls," *Transportation Research Record*, 1422, p. 29, Transportation Research Board, National Research Council, Washington, DC, 1993.

10-2 Eder, T.B., Bosscher, J.P., and Eldin, N.N., "Development of Engineering Criteria for Shredded Waste Tires in Highway Applications," Interim Report to Wisconsin Dept. of Transportation, University of Wisconsin-Madison, Madison, WI, 1990.

10-3 Environmental Study of the Use of Shredded Waste Tires for Roadway Subgrade Support, Groundwater and Solid Waste Division, Minnesota Pollution Control Agency, St. Paul, MN, 1990.

10-4 Read, J., Dodson, T., and Thomas, J., "Use of Shredded Tires for Lightweight Fill," Report No. DIFH-71-90-501-OR-11, Highway Division, Road Section, Oregon Department of Transportation, Salem, OR, 1991.

Scrap Tires: Disposal and Reuse

10-5 Humphrey, D.N., and Nickels, W.L., "Tire Chips as Subgrade Insulation and Light Fill," presented at the 18th Annual Meeting of the Asphalt Recycling and Reclaiming Association, February 23–26, 1994, Pompano Beach, FL.

10-6 Colorado Department of Transportation, 18550 Colfax, Aurora, CO 80011.

10-7 "A Report on the Use of Shredded Scrap Tires in On-Site Sewage Disposal Systems," prepared for the Department of Environmental Conservation, State of Vermont, May 22, 1990, Envirologic, Inc., 139 Main St., Brattleboro, VT 05301.

10-8 Hall, T., "Reuse of Shredded Waste Tires Material for Leachate Collection Systems at Municipal Solid Waste Landfills," Iowa Department of Natural Resources, Waste Management Authority Division, Grant Contract No. 90-655007, September 7, 1990, Shive-Battery Engineers and Architects, Inc., 1701 48th St., Suite 200 West, Des Moines, IA 50265-6755.

10-9 Humphrey, D.N., *Investigation of Exothermic Reaction in Tire Shred Fill Located on S.R. 100 in Ilwaco, Washington*, Humphrey, D.N., Consulting Engineer, P.O. Box 20, Palmyra, ME 04965, 1997.

10-10 Schlieder, G., "Temperature and Gas Monitoring in Shredded Tire Road Embankment—Spencer's Crest Drive, Eugene, Oregon," Gem Consulting Inc., P.O. Box 23635, Eugene, OR 97477-0147, prepared for J. Powell, Federal Highway Administration, August 9, 1996.

CHAPTER 11

Tire Pyrolysis

Description of the Process

One of the most widely discussed and studied methods of disposal of scrap tires is by pyrolysis or destructive distillation. The process consists of heating tires, or more conveniently tire chips, to temperatures above 315°C (600°F) in a confined space. At these temperatures, the tire rubber begins to decompose, generating an assortment of gaseous and liquid products, chiefly hydrocarbons. These products distill off and can be collected, leaving an involatile residue comprising the carbon blacks from the rubber compounds imbedded in carbonaceous residues from the decomposed rubber, and the steel wire segments of the tire, together with various inorganic materials. These include chiefly titanium dioxide pigment from the white sidewalls, silica, and perhaps some clay and talc.

The relative amounts of liquid and gas can be considerably influenced by the temperatures employed, but broadly we may expect to obtain one-third gas, one-third liquid, and one-third residue. Some of the gas will be required as process fuel for the process; the remainder, in principle, may be bottled and sold as liquefied petroleum gas (LPG). Similarly, the light liquid fraction can be sold as gasoline. A considerable volume of oil is collected, having a black color caused by carbon black particles that are transported as aerosol particles during distillation. The oil closely resembles No. 6 oil and could be used for the same purposes, including use as a rubber extender in tire manufacture. The steel can be collected magnetically and presumably sold as steel scrap. Likewise, the carbonaceous residue could be ground and sold as fuel.

Feasibility of Tire Pyrolysis

Technologically, at least, tire pyrolysis is feasible, and several prototype plants have been erected to demonstrate it. As early as 1974, the Goodyear Tire and Rubber Company began a tire pyrolysis project in conjunction with the Tire Oil Shale Company (TOSCO). In Rocky Flats, Colorado, using facilities designed to distill oil from Colorado oil shale, they pyrolyzed shredded tires. Technically, they were successful, but the project was abandoned because of the problems and expense of shredding tires at that time.[11-1] A whole-tire pyrolysis plant was operating for a time in Hamburg, Germany,[11-2] and a successful pyrolysis operation in Japan has been reported.[11-3]

Domestically, several prototype plants or pilot plants have been set up and have operated, at least sporadically, to demonstrate the process to potential entrepreneurs. However, according to the Scrap Tire Management Council, the number of tires pyrolyzed in 1996 was zero.[11-4]

When closely scrutinized, the economics of tire pyrolysis is forbidding for at least two reasons. First, if we pyrolyze a ton of tire chips into gas, oil, and carbon char that must be sold as fuel, the BTU level of our products is less than that of the tire chips themselves if burned as fuel because we had to expend considerable energy in the pyrolysis process. Likewise, the economic value of the products produced from that ton of tires is less than the value of the chips in today's market. Furthermore, to achieve this undesirable result, we had to make a big plant investment and incur heavy operations costs.

Prospectuses for tire pyrolysis business proposals often avoid this problem by showing the carbonaceous residue not as fuel, at coal prices, but as merchantable carbon black. The residue's value as carbon black is low partly because (as we have noted) it contains considerable inorganic materials as diluents. Tire companies do not use materials of such a low quality; therefore, the market would be severely limited at best. However, note that one company[11-5] has developed an auxiliary process to grind the residual char and separate the fine carbon black particles from the remainder of the mass. This material might achieve a limited market.

The second problem with the economics of tire pyrolysis proposals is that they assume plant operating costs typical of those in the petroleum industry. These are not even remotely applicable because the scale of operations is entirely different. Consider first, as we have shown previously, that scrap tires must be used close to the point of origin because of high transportation costs. In an urban metropolitan area having a population of two million, we might expect a similar number of scrap tires annually. If these scrap tires could be collected and delivered to the local pyrolysis plant, this would indicate a daily usage of 5,500 passenger tires. If we estimate that these scrap tires might generate 5 lb of oil per tire, we are talking about 27,000 gal. of oil

Scrap Tires: Disposal and Reuse

per day, or less than 1,000 petroleum barrels per day. This figure is below the size of most pilot plant operations in an oil refinery and can hardly achieve petroleum production economics. To put this in perspective, if 6 gal. of petroleum are needed to make one tire, then 500 gal. of gasoline are required to wear out that tire, assuming we are talking about a 50,000-mile tire and a 25-mpg vehicle. It is difficult to conceive of a location other than Metropolitan New York that could support an efficient pyrolysis plant in terms of the huge volume of tires required.

Additionally, a tire pyrolysis operation would have to be run with all of the safety precautions required in an oil refinery and presumably would have to be operated on an around-the-clock basis. A well-engineered pyrolysis operation in Struthers, Ohio, blew up during a Sunday shutdown and was never reopened. The plant in question[11-6] was built in 1982 by newly formed Carbon Oil and Gas in Struthers, Ohio, a suburb of Youngstown, on the site of a vacant steel-making plant. One of the founders, Morris O. Hill, the president and chief executive officer, was a well-qualified petroleum engineer with 27 years of experience at Standard Oil Company. A total of $3.5 million was invested over seven months to build the plant, which laid claim to being "this nation's first economically viable tire recycling and resource recovery system." The plant was designed to produce 7,500 gal. of oil per day. On start-up operation on January 6, 1983, the plant delivered 143 barrels of #6 oil to a broker. They encountered some problems in cooling the hot carbon char before discharge into open air and still lacked an OSHA clearance when, during a temporary shutdown on a Sunday morning, an explosion demolished the reactor. Presumably on cooling, air leaked back into the reactor and encountered hot petroleum gas, a constant risk in any petroleum refinery. The plant was never repaired nor reopened.

Considered broadly, the best thing that could be said for tire pyrolysis is that it provides a means to convert large amounts of scrap tires to useful by-products. However, pyrolysis could be profitable only if heavily subsidized (i.e., with a large and stable tipping fee). This seems unlikely in view of the more attractive alternative programs that are available today.

References

11-1 *Chemical and Engineering News*, June 10, 1974, p. 5.

11-2 *Chemical and Engineering News*, April 12, 1982, p. 53.

11-3 *Chemical Engineering*, December 31, 1979, p. 30.

11-4 "Scrap Tire Management Council: Use/Disposal Study, 1996 Update, April 1997," Scrap Tire Management Council, 1400 K St. N.W., Suite 900, Washington, DC 20005.

11-5 American Tire Reclamation, Inc., 6011 Joy Rd., Detroit, MI 48204.

11-6 *Rubber and Plastics News*, March 14, 1983.

Chapter 12

Other Solutions: Fishing Reefs and Molded Rubber-Plastic Blends

This chapter will consider two of the more innovative proposals for alleviation of the scrap tire problem. The first of these—fishing reefs—was an idea that arose at Goodyear Tire and Rubber Co. and was vigorously pursued and promoted by Goodyear. The second—molded rubber-plastic blends—is a much more recent development that is now coming on strong.

Scrap Tires in Fishing Reefs

One of the more unusual proposals to alleviate the scrap tire situation was to bundle the scrap tires together and sink them in relatively warm, shallow, coastal waters where they could serve as a refuge for minnows and fish fry from their predators. The objective was to increase the population of adult game fish.

Where it could be conveniently implemented, the idea was successful. First, an appropriate number of scrap tires had to be collected at a seaside location. Next, each tire had to be filled with a sufficient quantity of concrete mixture to overcome its buoyancy and to ensure that it would remain submerged. Following this, the tires were strapped into bundles with nylon tape and then hoisted onto the decks of vessels assembled for the purpose. Then the bundled tires were hauled to the designated location for the construction of the reef, where they were jettisoned in close proximity to each other.

Observation over a three-year period revealed that the tires in the bundles became encrusted with barnacles and other marine growth, which effectively cemented them together into large, stable blocks. Moreover, these scrap tire reefs served as effective havens for young fish as demonstrated by a discernible increase in adult game fish in subsequent seasons. Several successful reefs of this kind were produced in Florida, Maryland, and up to and including New Jersey. In colder, more northerly waters, the tropical marine growth does not materialize to the same degree, and movement of the bundles on the sand bottom produced by underwater currents slowly chafes the bindings and unbundles the tires.

By 1985, Goodyear had concluded that using scrap tires in fishing reefs was economically infeasible,[12-1] but local enthusiasm and volunteer sportsmen continued with projects in Ocean County in Maryland and in Cape May County in New Jersey, with the help of county subsidies.

Molded Rubber-Plastic Blends

Some of the best high-quality engineering plastics, such as ABS resins, are intimately mixed blends of high-softening thermoplastic polymer with small quantities of an appropriate rubber in a heterophase system. The rationale is to recognize that in a

Other Solutions

glassy material, such as the thermoplastic resin or even window glass, cracks are hard to initiate but propagate easily. The remedy then becomes to allow it to propagate only a short (microscopic) distance before it encounters a phase boundary and terminates and must then be reinitiated by another event. All this can be done in a finely designed and finely mixed, heterophase system in which one of the phases, in minor volume, is an elastomer gum with surface properties similar to the hard thermoplastic resin phase which is the predominant component. In the 1940s, pocket combs and children's toys from pure polystyrene or methyl methacrylate were hopelessly brittle and gave plastics a bad name. Blending with small amounts of a suitable rubbery polymer has largely eliminated these complaints, and rubber-plastic blends became commercially important in the 1940s as engineering plastics.

The possibility of using vulcanized crumb rubber in such compounds was not totally overlooked, and patents began to appear in the 1960s. However, activity was desultory at best, perhaps because of the unavailability of cheap crumb rubber. In the last few years, the picture has acquired a new dimension because of the heightened public interest in recycling, particularly of recycled plastics. Many of the more common plastics suitable for admixture with crumb rubber are identified with a molded-in identification symbol on the base of the object, and these objects are being source separated at the time of the trash collection in many towns and cities. Free access to this clean supply of sorted plastics represents a potential bonanza to plastics molders, who also have available to them at this time high-quality, clean, cheap crumb rubber. Innovative plastics extruders and molders are hastening to take advantage of the situation.

Scrap Tires: Disposal and Reuse

In Denver, Retek Materials Group, Inc.[12-2] has been making bump stops for parking lots for several years and is presently selling them to a major chain of consumer home improvement stores. (See Figure 12.1.) These bump stops are much lighter than the concrete ones being replaced, and they can be pigmented in attractive colors including blue for handicapped parking. Retek also makes large timbers for landscaping purposes. (See Figure 12.2.) In Canada, these and other plastic molded objects are being manufactured by Magnum Industries in Calgary,[12-3] including flat 1.5-in. plates for placement in flower beds.

In Michigan, Karl Loper[12-4] molds a 4-in. post with a flat plate on one end, suitable for a rural mailbox. (See Figure 12.3.) Similarly,

Figure 12.1. Parking-lot bump stops made from molded rubber-plastic blends. (Courtesy of Retek Materials Group, Inc.)

Other Solutions

Figure 12.2. Landscape timbers made from molded rubber-plastic blends. (Courtesy of Retek Materials Group, Inc.)

he molds longer posts for bird feeders and is prepared to mold house numbers on them. (See Figure 12.4.) The State of Michigan has tested and approved some of his posts for highway markers and has issued a purchase order for a trial quantity.

One of the major advantages expected from the use of molded rubber-plastic blends in these products is improved weatherability. Untreated wood posts rot after a few years. The ancient practice of creosoting wood posts is viewed as ecologically undesirable and is no longer permitted in some locations.

Scrap Tires: Disposal and Reuse

Fig. 12.3. A mailbox post and plate molded from polyethylene-crumb rubber blend.

In addition to materials suitable for rigid objects such as posts, we are seeing interesting properties in blends of relatively large amounts of ground scrap with smaller amounts of high-softening resin. Such blends seem to have properties of thermoplastic rubbers when properly processed. On cooling, the resin offers anchorage points for the rubber particles, maintaining them in fixed spatial relations in which they can continue to exhibit their rubbery properties in novel, useful molded articles and materials, as well as in mats and pads.

Scrap Tires: Disposal and Reuse

would be required to be branded by the manufacturer and sold with a deposit fee included in the sales price. Part of the deposit would be reclaimed by the owner when he turned in his worn-out tire. This proposal was renounced as unworkable by tire manufacturers and tire dealer organizations alike and soon was abandoned.

Minnesota, one of the early proponents of tire regulation, took a prudent early approach to the problem. The legislature funded a detailed study of the Minnesota scrap tire program and received an excellent analysis of the problem. The major piles of scrap tires were located, with most shown to be close to the Twin Cities (Minneapolis and St. Paul). One recommendation was to ensure that any depot for collection and confinement of scrap tires should be close to the Twin Cities to minimize transportation costs. However, when the legislature later took action, it chose to establish a scrap tire repository and tire chopping center in Babbitt, north of Duluth and 200 miles away from Minneapolis. Tire jockeys in the Twin Cities area soon realized that Wisconsin was only 20 miles away and had no restrictions on tire storage. Accordingly, large piles of Minnesota's scrap tires began to appear in rural, western Wisconsin and also in the eastern Dakotas. These adjoining states were obliged to pass their own scrap tire laws to repel out-of-state scrap tire disposal.

Similar situations arose in other places. Any state without regulations to control the storage and transportation of scrap tires was likely to receive scrap tires from a neighboring state or perhaps even a distant neighbor. When tipping fees in New York City reached $5 per tire, the scrap tires there were sometimes shipped profitably to scrap piles and landfills in the Midwest.

Chapter 13

Scrap Tire Regulations in the United States

Perception of a National Problem

Public recognition of our scrap tire problem in the United States began in the early 1980s. The situation came into full focus with the Winchester, Virginia, tire fire in 1983, which received national attention. (See Chapter 1 for more details on this fire.) The scrap tire situation was recognized as a national problem, but one that was more acute in some states than in others. It was perceived as an environmental problem and thus was presumably studied at length by the U.S. Environmental Protection Agency (EPA). Using its own proper criteria, EPA could not justify piles of scrap tires as being environmentally hazardous, regardless of the unsightly appearance of those tire piles and their potential as a fire hazard. In this situation, individual states (some with acute landfill problems) began to study the problem and to seek effective legislative remedies.

The Need for State Regulation

One of the early ideas to deal with the problem of scrap tires was to handle tires in the same way as bottles. In this solution, tires

Scrap Tires: Disposal and Reuse

The investment community is loath to get into scrap rubber products. Its expressed view is that the business overall is not as large as it should be. However, as we have noted elsewhere, the much greater available volume of thermoplastic resins could alter that point of view.

References

12-1 Personal communication, John Zimmer, Goodyear Tire and Rubber Co., 1144 E. Market St., Akron, OH 44316-0001.

12-2 Retek Materials Group, Inc., 5696 College Pl., Boulder, CO 80303.

12-3 Magnum Industries, 190 Hodsman, Regina, Saskatchewan, S4N5X4 Canada.

12-4 Karl Loper, 3170 Barnes Rd., Millington, MI 48746.

Other Solutions

Fig. 12.4. A molded post and plate for a birdfeeder, made from polyethylene-crumb rubber blend.

We can expect a surge in the production of molded and extruded objects from crumb rubber-plastic blends in the next few years. I believe that the major cause of the delay is the high investment cost for the necessary extrusion equipment, presses, and molds. A suitable 4-in. extruder would cost approximately $400,000, and even larger equipment might be necessary for optimal production speeds.

Not surprisingly, state legislatures became involved, and legislative activity has been heavy in recent years. Today, 48 states have significant regulations concerning the disposal of scrap tires.

Patterns Among State Regulations

Because the size and shape of the scrap tire problem in the United States are greatly influenced by demography, wide variations in the ways in which individual states deal with the problem could be expected. Nonetheless, a surprising amount of commonality exists among state regulations.

Most states have banned landfilling of whole tires, and a few have even banned landfilling of chopped tires. All scrap tires must be placed somewhere; therefore, provision must be made for their storage until they can be used. Accordingly, most states have specified adequate storage methods, often defining the maximum pile size and the need for surrounding berms or fences. They also usually prescribe the necessary recordkeeping on the origins of the tires accepted and the license provisions for the storage site, including fees.

To prevent the appearance of new unlicensed piles of scrap tires, most states have felt it necessary to license all haulers of scrap tires. Some states also require them to keep manifests of their deliveries.

Control of existing tire piles and the means by which they grow is the necessary first step and is apparently sufficient for some state programs at this time. Legislation in many other states provides for the elimination of offensive, existing piles of tires and often offers market incentives to scrap tire processors for the effective use of chopped tire products. This assistance may take the shape of grants or loans to buy tire chopping equipment for the processor, or direct subsidies to the users of the chopped tires.

For some time, Wisconsin has been willing to pay a subsidy of $20 per ton to users of Wisconsin scrap tires if the tires are used for fuel, and an additional subsidy of $20 per ton for other commercial uses of the tires. Likewise, Virginia now reimburses users of scrap tire products at $20 per ton. The response to these programs has been excellent.

Wisconsin recently modified its program to reward both the processor and the users of scrap tire products. The processor, who chops the tires, can collect a subsidy of $40 per ton; the user can do the same. The scrap tires must have originated in Wisconsin, but payments have been made to processors outside Wisconsin.

Funding

In recent years, state budgets have received significant pressure, and the public temper has favored budget cutting rather than new and expensive programs. Accordingly, one of the major obstacles in achieving effective tire regulation has been the question of funding: How do we fund tire regulations without an obvious increase in the general tax rate and avoid organized opposition from strong lobbying groups? A second requirement was that the required money would have to be collected through an existing channel rather than a new collection mechanism, and another organization would have to be staffed.

In its program, Minnesota elected a $4 charge on vehicle title transfers. This generates approximately $4 million per year, of which two-thirds has gone into the stockpile clean-up program. The remainder has gone into grant and loan programs for scrap tire recyclers and users.

Wisconsin chose to put a $2-per-tire tax on new vehicles (including 18-wheelers) to generate approximately $3 million per year. Likewise, Michigan has a $0.50-per-tire disposal surcharge on vehicle registration, and New Mexico has added a charge of $1 per tire to the vehicle registration fee.

The most popular funding device has been a simple increase of $1 or $2 on the retail sales tax on a tire. In some states, that tax is only $0.50. California remains different, charging a $0.25 fee on all worn-out tires left with a tire dealer, generating approximately $3 million per year.

New Programs

Scrap Tire News publishes an excellent biennial report on state scrap tire management programs,[13-1] summarizing the significant details of the programs in each state. The publication also lists the appropriate contact for more detailed information. Anyone contemplating scrap tire activity in his own or any other state should consult this reference.

State legislative activity remains high. John Serumgard reports[13-2] that in the first six months of 1995, nearly 100 new bills were introduced in 30 states. These bills involved modification of existing programs rather than comprehensive new legislation.

Scrap Tires: Disposal and Reuse

In conclusion, note that the various states have enacted programs to suit their own needs, and these have proven satisfactory to date. However, some believe that scrap tires are a national problem in which the federal government should play a role. To this end, Representative Esteban Torres of California has proposed a federal statute several times, which would shift the entire responsibility for scrap tire disposal to the tire companies that produced the tires, in proportion to their sales. For various reasons, his bills have met with insufficient support, if not apathy.

References

13-1 "Scrap Tire News Special Report 1997," Recycling Research Institute, 133 Mountain Rd., Suffield, CT 06078.

13-2 "State Legislation: A Mid Year Review," *Scrap Tire News*, Vol 9, No. 9, 1995.

CHAPTER 14

Overview and Projections for the Future

Overview of Current Usage

The foregoing chapters clearly demonstrate that the scrap tire problem in the United States is being vigorously and successfully addressed. The Scrap Tire Management Council has been conducting biennial surveys of scrap tire disposal. The 1996 update of its scrap tire use/disposal study is the most comprehensive compendium of information available on this topic.[14-1] Table 14-1 summarizes one of the tables in the report.

As in the previous issue, the Scrap Tire Management Council report reappraises the number of scrap tires to be found in stockpiles existing in various states, both legally and illegally, and concludes that a conservative estimate of 800 million is still the best answer, due in part to the progress that many states, especially Illinois, Wisconsin, Minnesota, Maryland, Florida, and Oregon, have made in eliminating such piles. On this basis, at some time in the latter part of 1998, we shall be reusing scrap tires at the same rate at which they are being generated. Further increases in reuse must necessarily involve the existing piles stored outdoors.

TABLE 14-1
REPORTED AND ESTIMATED DEMAND FOR SCRAP TIRES BY MARKET SEGMENT
(MILLIONS OF TIRES)

Market Segment	1997 Estimate	1998 Estimate
Tire Derived Fuel (TDF):		
Cement Kilns	53	58
Pulp and Paper Mills	37	39
Utility Boilers	32	36
Dedicated Tire to Energy	15	16
Industrial Boilers	23	25
Resource Recovery Facilities	8	10
Lime Kilns	2	3
Copper Smelters	1	1
Iron Cupola Foundries	<u>1</u>	<u>4</u>
Subtotal Fuel	172	186
Products:		
Size Reduced Rubber	15	18
C/S/P Products	8	8
Civil Engineering	14	18
Agricultural	2.5	2.5
Export	15	15
Miscellaneous Uses	<u>1.5</u>	<u>1.5</u>
Total	228	249
Annual Generation	270	275
Scrap Tire Markets as Percent of Total Tires Generated	84	90

Overview and Projections for the Future

Although obviously true, this statement does not throw much light on the remaining problems but rather conceals from us many things of importance.

Let us consider the situation from a different viewpoint. Actually, there is good reason to believe that our rate of reuse of scrap tires will continue to grow for some time. Consider that during the past decade, there has been significant interest in this problem, and a large amount of money has been invested in its solution. Some of it has been illspent and lost. But much of it has been spent well, and now scrap tire processors in every state have learned how to compete and intend to reward themselves with the solutions to this problem. A great momentum exists here, which is not easily stopped. There are many cement kilns burning tires. Although many do not yet burn tires, some will become converts if real value exists. There is no reason to think that we have saturated the market for TDF, and the effort that has gone into this will continue to bear fruit. Likewise, the effort that has gone into rubber asphalt blends also will pay off in terms of increased value and adoptions. The potential value of tire chips in civil engineering applications is already too high to ignore. The problems that exist certainly are solvable, given the potential value of the solutions. They are already beginning to ship tire chips (Crown III) to golf courses in Montana, and you could argue that Montana never had a scrap tire problem.

Behind all these new development ideas for scrap tires are many attractive possibilities that have not even been investigated yet. For example, in the Northeast, we shall backfill with ground rubber some basements in newly constructed homes to prevent heat loss. In northern climates, winter heat loss through basement walls and floors is approximately 2 BTU/ft^2/hr. The first few people to do this probably will not recover their costs. However, if widely adopted, the fuel savings and reduction in use of fossil fuels could be enormous and at trivial cost. In addition, such treatment should radon-proof basements because tire chips would be a much more effective barrier to radon than the polyethylene film currently used.

Again, in previous chapters, some public concern has been reflected about leachate problems from scrap tire particles. What has not been mentioned is that tire rubber itself is an effective solvent for many of the more hazardous environmental pollutants, such as PCBs and dioxin, and could be useful in their control. Professor Tuncil B. Eder of the University of Wisconsin has observed that ground tires mixed into clay used in landfill barriers can, through absorption, impede migration of some of these pollutants through the clay retaining wall and extend the protective life of the landfill.[14-2]

Conclusions and Projections

We are rapidly approaching a complete solution to our scrap tire problem in the United States. At the point where we are using scrap tires as quickly as they are produced, the tires in the stockpiles will acquire some value and will be used, chiefly as fuel. Now we can address the interesting question of what happens when that scenario occurs. The vast amount of effort that has been going into innovative solutions to the problem has a kind of momentum and will not suddenly cease; rather, it will continue to bear fruit with new products from scrap tires, which will compete in the market for a shrinking supply. What happens then? Many different scenarios can be written, but they all must begin with a consideration of values.

Overview and Projections for the Future

Some cement kilns are being paid to burn tires and thus have fuel with a negative cost. Likewise, some industrial boilers receive tire chips free of charge; others receive tire chips at costs far less than the cost of chopping the tire. In those cases, the users are participating in the tipping fee. When the suppliers can find alternative markets, these practices must cease as the tires or tire chips begin to assume some true economic value. When used as fuel, the economic value of tire chips relates to the alternative fuel, usually coal. Coal is not cheap. Close to the mine, stoker coal might cost as little as $25/ton, but shipping costs can easily double the cost. On the other hand, tire chips would tend to be generated locally, thereby avoiding high freight costs. If tire chips can be sold at or close to their true economic value, chopping tires could become profitable if we reckon the true cost of chopping those tires to be approximately $30/ton.

Fuel use probably presents the lowest economic value for tire chips. Close to fuel use would be some civil engineering uses. Tires used as a roadbase or as a light aggregate backfill can be produced inexpensively and because of the inherent value are likely to be widely used in the future. Maine is beginning to specify tire chips for particular jobs. One such job projected for 1997 will require a million or more chopped tires, at an estimated savings of $300,000 over the competitive material (foamed polystyrene).

For many civil engineering applications, the cost of rubber chips could be favorable compared to gravel because of the higher density and lower volume of gravel and the higher shipping costs associated with its use.

Scrap Tires: Disposal and Reuse

For many of the other uses we have discussed in this book, such as soil amendment, mats, playground materials, and molded objects, the value of ground rubber is approximately $100/ton or higher. As these uses proliferate, they will tend to replace fuel uses.

The value of rubber in asphalt extension is established in those states in which the practice has gained wide acceptance. For such uses, the production cost of the finely ground rubber is significantly higher, as is the selling price of approximately $200/ton. If the level of use originally proposed by ISTEA was in effect, this would become the dominant use and probably would seriously impinge on the availability as low-cost fuel. The repeal of ISTEA has deferred this possibility but has not necessarily interfered with the successful issue of this result as development and testing continue.

As the scene changes as suggested above, repercussions will be realized on the tipping fees. In general, tire dealers do not like to pay tipping fees of approximately $1 per tire or more. Today, many scrap tire processors are scrambling to make a living, and great overcapacity exists to make chips smaller than 1 in. However, as they become profitable and greater competition for scrap tires arises, tipping fees will decrease. Furthermore, it is not too wild a dream to think that as the value of a scrap tire rises to its full material value, it will exceed the tipping fee, and tire dealers may again begin to sell scrap tires.

Major industrial developments involving novel materials can be expected to take time to mature, and such has been the case with scrap tire usage. However, within the time it has taken to publish this book, we begin to see wide-scale adoption of some of the

Overview and Projections for the Future

things we have discussed and also the appearance of other new products.

The use of tire chips in equestrian arenas is rising sharply, and many scrap tire processors are participating. Norman Emanuel has filled one 10,000-ton order for use in Great Britain and is completing shipment of a second order in late 1997. This 10,000 tons of 1-inch rubber chips involves more than a million scrap tires.

The *1997 Scrap Tire and Rubber Users Directory* lists several dozen manufacturers of mats. To my knowledge, at least one of these manufacturers uses 2 million pounds of crumb rubber per month.

The use of rubber chips in septic systems is beginning to catch on, notably in the Carolinas and Georgia. Perhaps this is occurring because gravel is lacking in these states, and crushed traprock is expensive. Likewise, the use of rubber chips in landfills continues to grow in large volumes in other areas.

Rubber chips embedded in concrete make good sound-dampening berms for blocking highway noise. This use is beginning to consume substantial quantities of rubber chips. The State of Maine continues to use scrap tire pieces in highway construction, and each job there uses a million tires or more.

At the other end of the spectrum, one orchid grower has found rubber chips to be a superior medium in which to grow orchids. This information appeared in a publication for orchid growers and produced a rash of inquiries to Jai-Tire Industries, which had provided the appropriate size of tire chips.

The existing piles of scrap tires are beginning to provide the rubber for some of these markets. However, with the rising uses we are seeing, these piles may soon be depleted—I estimate that they cannot last three years. At that point, the whole scene could change, and tipping fees would begin to erode as the intrinsic material value of scrap tires improves. The collectors of scrap tires would then begin to play a major role because they would control the supply of scrap tires to the processors. During this period of perhaps three years, we may even be able to conclude that the national scrap tire problem in the United States has been solved.

References

14-1 "Scrap Tire Use/Disposal Study 1996 Update," Scrap Tire Management Council, 1400 K St. NW, Washington, DC 20005, April 1997, 714-2.

14-2 Park, J.K., Kim, J.V., and Eder, T.B., "Mitigation of Organic Compound Movement in Landfills by Shredded Tire," Environmental Geotechnics Report No. 95-2, Department of Civil and Environmental Engineering, University of Wisconsin, Madison, WI 53706, May 10, 1995.

Index

Agrimats, 73-74
   see also Mats
Ambient grinding, 34-35, 35*f*
Asphalt
   description of, 59-60
   rubber in, 57-60

Backfill, tire chips as, 98
Biological oxidation of rubber, fire and, 105
Birdfeeder posts and plates, of molded rubber-plastic blends, 116-117, 119*f*
Blade sharpening, for hook and shear shredders, 23
Boiler. *See* Steam boiler
Boiler fuels, whole tires as, 50-51

Carbon Oil and Gas pyrolysis operation, 110
Chopping tires. *See* Tire chopping
Coal fuel supplement, TDF as, 51-52
Collection process
   dealer-jockey relationship and, 10-14, 13*f*
   disposition of medium-truck tires, 14-15
   role of tire jockey in, 9-10
Composting organic materials, fire and, 104
Concrete, rubber chips embedded in, 133
Costs
   of crumb rubber, 63-64
   of equipment, 27-28

Costs *(continued)*
    of shredders, 27-28
    in utilization of tire chips or crumb rubber, 41-42
Cowmats, 73
Crown III, 92
Crumb rubber
    in crack sealants, 61
    on golf courses, 89, 90*f*, 91-92
    in highway construction, ISTEA mandate on, 65-67
    mats from, 72*f*
    in paving courses, 62-65
    in repair membranes, 61-62
    as soil amendment, 87-89, 90*f*, 91-92
        effects on soil, 86-87
    see also Rubber
Crumb rubber-plastic blends, 114-120
    birdfeeder posts and plates from, 116-117, 119*f*
    landscape timbers from, 117*f*
    mailbox posts and plates from, 116-117, 118*f*
    parking-lot bump stops from, 116*f*
Cryogenic grinding, 35-37

Dealer-jockey relationship, 10-14
Destructive distillation. *See* Tire pyrolysis
Diamond Z hammermill, 25-26, 26*f*
Disposition of medium-truck tires, 14-15
Dry process of crumb rubber and asphalt, 64-65

Embankment fire, 101
Engineering properties of rubber particulates, 40*t*
Equestrian arenas, rubber chips used in, 78-80, 79*f*, 133
Eugene, Oregon, shear slope failure in, 103

Index

Falling Springs Road, Washington, embankment fire in, 101
Fill, tire chips as lightweight aggregate in, 95-96, 97*f*
Fire hazards
    tires as, 2-3
    see also Tire fires
Fires
    in embankment of tire chips, 101
    organic materials composting and, 104
    oxidation of steel to iron oxides and, 104
    possible explanations for, 104-105
    as problem, in highway construction, 100-102
    in retaining wall, 100
    in road fill, 101-102
    see also Tire fires
Fishing reefs, scrap tires in, 113-114
Flammability of tires, 2
Fuels
    boiler, whole tires as, 50-51
    coal, supplemental, TDF as, 51-52
    heat of combustion of, 48*t*
    tire pyrolysis and, 108
    tires burned as
        characteristics of, 47-50
        heat of combustion of various fuels, 48*t*
Funding
    effective tire regulation and, 124-125

Gasoline, tire pyrolysis and, 107-108
Glenwood Canyon, Colorado, retaining wall fire in, 100
Golf courses, rubber chips as soil amendment on, 89, 90*f*
Goodyear Tire and Rubber Company, tire pyrolysis project of, 108
Granulators, 31, 32*f*
Gravel-surfaced road, tire chips as subfill insulation beneath, 98

Grinding
    ambient, 34-35, 35*f*
    cryogenic, 35-37
    wet, 37
Gypsy tire jockey, 12

Hagersville, Ontario, tire fire in, 3
Hammermills, 25-26
    Diamond Z, 26*f*
Hazards. *See* Fire hazards; Health hazards
Health hazards, from scrap tires, 3-4
Highway construction
    ISTEA mandate on use of crumb rubber in, 65-67
    problems in using tire chips in, 100-105
        embankment fire, 101
        retaining wall fire, 100
        road fill fire, 101-102
        shear slope failure, 103
Holman shredder, 23-24, 24*f*, 25*f*
Hook and shear shredders, 21*f*, 21-22
Hot spots, in highway construction, 100-102

Ilwaco, Washington, road fill fire in, 101-102
Impact attenuation of surface systems under and around playground equipment, 75
Insulation. *See* Subgrade thermal insulation
Intermodal Surface Transportation Efficiency Act (ISTEA)
    mandate on use of crumb rubber in highway construction, 65-67
ISTEA. *See* Intermodal Surface Transportation Efficiency Act

Landfills, tire chips in, 98-99
Landscape timbers, from molded rubber-plastic blends, 116, 117*f*
Landslides, correction of, 96

Index

MacAdam, John, asphalt and, 57
Mailbox posts and plates, from molded rubber-plastic blends, 116-117, 118*f*
Maine. *See* North Yarmouth, Maine
Mats
    from crumb rubber, 72*f*
    see also Agrimats; Rubber mats
Mayer, Gummi, 50
McDonald process, 64
Medium-truck tires, disposition of, 14-15
Minnesota, scrap tire regulations in, 122
Modesto, California, tire pile in, 12, 13*f*
Molded rubber-plastic blends, 114-120
    birdfeeder posts and plates, 116-117, 119*f*
    landscape timbers, 116, 117*f*
    mailbox posts and plates, 116-117, 118*f*
    parking-lot bump stops, 116, 116*f*
Mosquito infestation, tire piles and, 3-4

National problem of scrap tires, perception of, 121
North Yarmouth, Maine, tire chips as fill beneath paved road in, 96, 97*f*

Oil, tire pyrolysis and, 108
Orchids, rubber chips and, 133
Oregon Department of Transportation, shredded tires to correct landslide used by, 96
Organic materials composting, fire and, 104
Oxidation of steel to iron oxides, fire and, 104

Parking-lot bump stops, from molded rubber-plastic blends, 116, 116*f*
Playground surfaces
    rubber chips as, 77*f*
    rubber mats for, 74-75

Playturf, from scrap tires, 76-78, 77*f*
Polyethylene, as problem, 42-43
Polyethylene-crumb rubber blend
    birdfeeder posts and plates from, 116-117, 119*f*
    mailbox posts and plates from, 116-117, 118*f*
Polypropylene, as problem, 42-43
Portland cement industry, TDF in, 53-55
Problems
    in highway construction, 100-105
    when road is built on hillside, 96
    of scrap tires, 42-43
        description, 1
        fire hazard, 2-3
        health hazards, 3-4
        national situation, 121
    with tire cord and tire wire, 32-34
    see also Scrap tire problems
Pyrolysis, 2
    see also Tire pyrolysis

Railroad crossings, rubber in, 67-69, 69*f*
Rat infestation of tire piles, 3
Rebound, 87-89
Recycling Research Institute, 26
Refiner mill, 35*f*
Retaining wall fire, 100
Retaining walls, tire chips as backfill behind, 98
Road fill, fire in, 101-102
Road-building material, rubber in, 57
Rotary shear shredder
    with detachable knives, 25*f*
    with one-piece counter-rotating knives, 21*f*, 21-22
Rubber
    in asphalt, 57-60
    biological oxidation of, 105

Index

Rubber *(continued)*
    in equestrian arenas, 78-80, 79f, 133
    in railroad crossings, 67-69, 69f
    see also Crumb rubber
Rubber chips
    embedded in concrete, 133
    in equestrian arenas, 78-80, 79f, 133
    as protective playground surface, 77f
Rubber fines, fire and, 105
Rubber mats
    for playground surfaces, 74-75
    uses of scrap tires in, 71, 72f, 73
    see also Mats
Rubber particulates, engineering properties of, 40t
Rubber railroad crossings, 69f
    manufacturers of, 68
Rubber-plastic blends. *See* Molded rubber-plastic blends
Rubber-polyurethane mats, 75

Safety precautions, in tire pyrolysis, 110
SAM. *See* Stress Absorption Membrane
SAMI. *See* Stress Modifying Membrane Interlayer
Schlieder, Gunnar, 103
Scrap Tire Users Directory, 26
Scrap tires
    collection of
        dealer-jockey relationship, 10-14
        disposition of medium-truck tires, 14-15
        tire jockey, 9-10
        tire pile in Modesto, California, 13f
    conclusions and projections on, 130-134
    fuel uses of
        characteristics of tires burned as fuel, 47-50
        heat of combustion of various fuels, 48t
        TDF in portland cement industry, 53-55

Scrap tires *(continued)*
    fuel uses of *(continued)*
        TDF as supplemental coal fuel, 51-52
        TDF as wood supplement, 52-53
        whole tires as boiler fuels, 50-51
    management programs for, 125-126
    in molded rubber-plastic blends, 114-120
        birdfeeder posts and plates, 116-117, 119*f*
        landscape timbers, 116, 117*f*
        mailbox posts and plates, 116-117, 118*f*
        parking-lot bump stops, 116, 116*f*
    origins of, 4-8, 6*t*
    problem of, 42-43
        description of, 1
        dual nature of, 15-17
        fire hazard, 2-3
        health hazards, 3-4
        national, 121
    properties of, 39-41, 40*t*
    reported and estimated demand for, 128*t*
    regulation of
        funding and, 124-125
        need for, 121-123
        new programs, 125-126
        patterns among, 123-124
    reuse of, overview of current usage, 127-130, 128*t*
    in sewage sludge composting, 81-82, 83*f*
        tire chips replace wood chips, 82, 84-86
        West Windsor Water Pollution Control Plant, 83*f*
    in soil amendments
        crumb rubber, 87-89, 90*f*, 91-92
        effects of crumb rubber on soil, 86-87
    source separation of, 7
    transportation uses of
        crumb rubber in crack sealants, 61
        crumb rubber in paving courses, 62-65

Index

Scrap tires *(continued)*
    transportation uses of *(continued)*
        crumb rubber in repair membranes, 61-62
        ISTEA mandate on use of crumb rubber in highway construction, 65-67
    uses for
        agrimats, 73-74
        in asphalt, 57-60
        in equestrian arenas, 78-80, 79*f*, 133
        in fishing reefs, 113-114
        as fuel supplement, 47-55
        hierarchy of, 44-45
        in molded rubber-plastic blends, 114-120
        playground surfaces, 74-75
        playturf, 76-78, 77*f*
        in railroad crossings, 67-69, 69*f*
        rubber mats, 71, 72*f*, 73
        in sewage sludge composting, 81-86
        in soil amendments, 86-92
    see also Tires
Screening and sorting, of tire chips, 28-29
Septic fields, tire chips in, 99-100
Septic systems, use of rubber chips in, 133
Sewage sludge composting facilities, 82, 83*f*
Sewage treatment plant sludge, 81-82
    composting, 83*f*
        tire chips replace wood chips in, 82, 84-86
        West Windsor Pollution Control Plant facility, 83*f*
Shear slope failure, in Eugene, Oregon, 103
Shredded tires, to correct landslide under highway embankment, 96
Shredders
    cost of, 27-28
    hammermills, 25-26, 26*f*
    Holman shredder, 23-24, 24*f*, 25*f*
    hook and shear, 21*f*, 21-22

Shredders *(continued)*
    rotary shear shredders
        with detachable knives, 25*f*
        with one-piece counter-rotating knives, 21*f*, 21-22
    single-rotor machines, 24-25
    Untha tire grinder and shredder, 27*f*, 27-28
Single-rotor shredders, 24-25
Soil, effects of crumb rubber, 86-87
Soil amendment, crumb rubber as, 87-89, 90*f*, 91-92
Sorting. *See* Screening and sorting
Source separation, 7
State regulations
    need for, 121-123
    patterns among, 123-124
Steam boilers
    fueled with whole tires, 50-51
    see also Boilers
Stoker-fired boiler, 51-52
Stress Absorption Membrane (SAM), 61-62
Stress Modifying Membrane Interlayer (SAMI), 62
Struthers, Ohio, Carbon Oil and Gas pyrolysis operation in, 110
Subfill insulation. *See* Subgrade thermal insulation
Subgrade thermal insulation, tire chips as, 98

TDF. *See* Tire derived fuel
Temperature monitoring equipment
    tire chips in highway construction and, 103
Tire chips
    as backfill behind retaining walls, 98
    coarsely chopped, utility of, 29-30
    comminution of
        ambient grinding, 34-35
        cryogenic grinding, 35-37
        granulator mechanism, 32*f*
        particle size related to process, 31

Index

Tire chips *(continued)*
    comminution of *(continued)*
        refiner mill, 35*f*
        tire cord and tire wire problems, 32-34
        wet grinding, 37
    engineering properties and value of
        cost considerations, 41-42
        hierarchy of uses for scrap tires, 44-45
        perception of problem, 42-43
        properties of rubber particulates, 40*t*
        properties of scrap tires, 39-41
    as fill beneath paved road, 96, 97*f*
    fire problems with, 100-102
    highway construction problems, 100-105
    intermediate rechopping of, 31, 32*f*
    in landfills, 98-99
    as lightweight aggregate in fill, 95-96, 97*f*
    screening and sorting of, 28-29
    in septic fields, 99-100
    as subgrade thermal insulation, 98
    wood chips replaced by, 82, 83*f*, 84-86
Tire chopping
    abrasive tire rubber compounds and, 20
    difficulties in, 19-20
    primary, 20-28
        detail of Holman shredder, 24*f*
        Diamond Z hammermill in operation, 26*f*
        rotary shear shredder with detachable knives, 25*f*
        rotary shear shredder with one-piece counter-rotating knives, 21*f*
        Untha tire grinder and shredder, 27*f*
    rubber and wire resist cutting in, 19-20
    screening and sorting tire chips, 28-29
    utility of coarsely chopped tire chips, 29-30
Tire cord problems, 32-33

Tire derived fuel (TDF)
  in portland cement industry, 53-55
  as supplemental coal fuel, 51-52
  as wood supplement, 52-53
Tire fires
  in Hagersville, Ontario, 1990, 3
  in Winchester, Virginia, 1983, 2-3
Tire jockey
  dealer-jockey relationship, 10-14
  role of, 9-10
Tire piles
  in Modesto, California, 13*f*
  mosquito infestation of, 3-4
  rat infestation of, 3
Tire processing
  chopping
    blade sharpening, 23
    cost of equipment, 27-28
    Diamond Z hammermill in operation, 26*f*
    difficulties in, 19-20
    hammermills, 25-26, 26*f*
    Holman shredder, 23-24, 24*f*, 25*f*
    hook and shear shredders, 21*f*, 21-22
    primary, 20-28
    rotary shear shredder with detachable knives, 25*f*
    rotary shear shredder with one-piece counter-rotating knives, 21*f*
    screening and sorting tire chips, 28-29
    Untha tire grinder and shredder, 27*f*
    utility of coarsely chopped tire chips, 29-30
  comminution of tire chips
    ambient grinding, 34-35, 35*f*
    cryogenic grinding, 35-37
    granulator mechanism, 32*f*
    particle size and, 31-32, 32*f*
    refiner mill, 35*f*

Tire processing *(continued)*
    comminution of tire chips *(continued)*
        tire cord and tire wire problems, 32-34
        wet grinding, 37
Tire pyrolysis
    Carbon Oil and Gas and, 110
    description of, 107-108
    economics of, 108-109, 109-110
    feasibility of, 108-111
    prospectuses for, 109
Tire retreading, 5
Tire shipments, domestic, 6*t*
Tire wire problems, 33-34
Tires
    abrasive rubber compounds in, 20
    burned as fuel, 47-50, 48*t*
    as fire hazard, 2-3
    rubber and wire of, 19-20
    whole, as boiler fuels, 50-51
    see also Scrap tires
Transportation uses of scrap tires
    crumb rubber
        in crack sealants, 51
        ISTEA mandate on use of, 65-67
        in paving courses, 62-65
        in repair membranes, 61-62
    rubber
        in asphalt, 57-60
        in railroad crossings, 67-69, 69*f*

Untha tire grinder and shredder, 27*f*, 27-28

West Windsor Water Pollution Control Plant, Ontario, Canada, 83*f*
Wet grinding, 37

Wet process of crumb rubber and asphalt, 64
Whole tires. *See* Tires
Winchester, Virginia, tire fire in, 2-3
Wisconsin, state regulations in, 124
Wood supplement, TDF as, 52-53
World War II, use of rubber in, 58-59
Worn-out tire, definition of, 10

Zement, Dyckerhoff, 54-55

About the Author

Robert H. Snyder has more than 50 years of experience in rubber and tire development. Born in Montana in 1918, Dr. Snyder earned a B.S. in chemistry at the University of Michigan in 1940. After two years as a junior chemist at Hoffman-LaRoche, he went to work in 1942 for what was then the United States Rubber Company at its General Laboratory in Passaic, New Jersey. Dr. Snyder continued working for that company or its successor for 45 years, except for a three-year leave of absence. During this leave of absence, he attended the University of Chicago and earned his Ph.D. in organic chemistry in 1948. Dr. Snyder then returned to the United States Rubber Company General Laboratory to lead research teams in organic chemical synthesis and vinyl polymerization, and he became Director of Synthetic Rubber Research. In 1995, Dr. Snyder transferred to the company's Tire Division in Detroit, Michigan, as Director of Materials Research. He continued to work there in rubber and tire development until retiring in 1987 as Vice President of Tire Technology.

Dr. Snyder has written more than 20 published papers and holds 36 patents. During his career, he was active in many technical organizations within the industry, including directorship in the Rubber Division of the American Chemical Society, chairmanship of the Detroit Rubber Group, and, for 10 years, chairmanship of the Highway Tire Committee of the Society of Automotive Engineers.

Dr. Snyder's interest in tire recycling began many years before his retirement. However, after retirement, the topic has become his dominant activity, with participation as a speaker at recycling symposia and as a consultant. Most recently, at the 1997 Meeting of the International Tire and Rubber Association in Louisville, Dr. Snyder received an Industry Pioneer Award for his innovative contributions to scrap tire recycling.